Non-Linear Pulses in Integrated and Waveguide Optics

Non-Linear Pulses in Integrated and Waveguide Optics

A.B. SHVARTSBURG
*Professor, Central Design Bureau for Unique Instrumentation,
Russian Academy of Sciences*

OXFORD NEW YORK TOKYO
OXFORD UNIVERSITY PRESS
1993

Oxford University Press, Walton Street, Oxford OX2 6DP

Oxford   New York   Toronto
Delhi   Bombay   Calcutta   Madras   Karachi
Kuala Lumpur   Singapore   Hong Kong   Tokyo
Nairobi   Dar es Salaam   Cape Town
Melbourne   Auckland   Madrid
and associated companies in
Berlin   Ibadan

Oxford is a trade mark of Oxford University Press

Published in the United States
by Oxford University Press Inc., New York

A catalogue record for this book is available from the British Library

Library of Congress Cataloging in Publication Data
Schvartsburg, A. B.
 Non-linear pulses in integrated and waveguide optics / A. B.
Schvartsburg.
  (Oxford series in optical and imaging sciences ; 3)
  Includes bibliographical references and index.
  1. Integrated optics. 2. Optical wave guides. 3. Nonlinear
optics. I. Title. II. Series.
TA1660.S38   1993   *92–42889*

ISBN 0–19–856358–2

Typeset by Colset Pte Ltd., Singapore
Printed in Great Britain on acid-free paper by
Biddles Ltd., Guildford & King's Lynn

Preface

This book may be considered as one of the first attempts at a self-contained introduction to a new field of applied physics on the frontier between integrated optics and non-linear wave dynamics. The results obtained in this field have been dispersed hitherto in a series of journals devoted to quantum electronics, optics, instrumentation, and IR physics. This work appeared at the outset on a case-by-case basis but manifested the appearance of common approaches and methods. The analysis of overlapping problems in these publications permits one to indicate the penetration of non-linear concepts from pure theory to the innovative elaboration of sophisticated devices for light wave technology, signals processing, and all-optical information systems. The rapid development of this tendency leads to formation of the physical fundamentals of non-linear integrated optics and non-linear communication technique that are considered to be the 'hot' topics of modern optoelectronics. These topics are in demand for science, commerce, and defence.

The development of optoelectronics was based firstly on using new types of transmitters (laser, photodiode) and receivers (photodetector) in combination with the power-independent optical devices for coupling, modulation, and switching, such as the well-known directional couplers, Bragg filters, and birefringent waveguides. The main concepts of these devices were borrowed from traditional 'linear' microwave techniques being power-independent, the parameters of such optical elements could not be tuned during the operating processes. However, the high power density of radiation trapped in the waveguides stimulated interest in non-linear tunable all-optical systems. These efforts gave rise to the conversion of linear devices, e.g. for coupling and switching, into the energy-dependent ones using non-linear materials in the waveguiding regions. On the other hand these tendencies stimulated the design of purely non-linear circuits, based, for instance, on bistability or induced non-reciprocity, which have no counterpart in linear integrated optics. Both of these fields are now entering a new era, in which the novel results of research into light wave interactions are being used.

This book is intended to bridge the gap between the novel results in non-linear optics and their use in the design of wide classes of all-optical circuits for particular applications. The following are the main goals of this book:

1. To visualize the physical foundations of tunable powerful optoelectronic systems and to develop standard unified methods of analysis for these systems.
2. To elaborate methods to optimize the contributions of power-independent and power-dependent effects in the dynamics of radiation fluxes in guiding systems.
3. To explain the concepts of generalized non-linear regimes in integrated optics, which naturally involve the well-known linear regimes as their particular cases.

This book is centred mainly on third-order non-linear optical phenomena in layered structures, guiding interfaces, and channel waveguides. The induced non-linear polarization in these phenomena depends on the product of three wave fields. The following key concepts, inspired by consideration of the above-mentioned goals in the framework of third-order effects, run throughout this book.

1. Non-linear spatial oscillations of amplitudes and phases in the course of propagation of travelling modes due to interactions between their polarization components. Unlike the fixed amplitude-phase structure of modes given by trivial low-intensity limits of waveguide theory, this type of evolution is a vital factor for controlled tuning of trapped radiation.
2. Self-, cross-, and induced-phase modulations of guided modes. These non-linear effects make the interference power-dependent providing ultrafast re-building of light fluxes.
3. Sensitivity of non-linear dynamics of trapped wave fields to the amplitude, phase, and spectral envelopes at the waveguide input. This sensitivity is important for programmed shaping of pulses and sensing processes.
4. Non-reciprocity of phase shifts of overlapping strong light beams. The variety of these effects for co- and counter-propagating beams, depending upon the waveguide geometry, makes an influence on the spatial scales and depth of wave–wave cross-modulation in numerous integrated optics devices.
5. Non-linear multi-layer interfaces for tuning of reflection and refraction of laser beams. Such interfaces exhibit the characteristics of broad-banded cavityless amplitude–phase bistability of tuned beams.

This book consists of three parts: Chapters 1 and 3 are devoted to CW and pulse energy flows in non-linear guiding systems. Chapter 2 is concerned with tuning of radiation flows at a non-linear interface. The physical foundations of the phenomena analysed in each chapter are outlined in the Introduction to that chapter.

Non-linear waveguide circuits in integrated optics are analysed in Chapter 1. They are usually short, and thus dispersive effects can not accumulate. The ideas of controlled tuning of trapped modes based on periodicity, non-reciprocity, and bistability of non-linear optical circuits are illustrated by means of numerous waveguide devices. Many integrated optics applications use rectangular strip waveguides which are embedded in other dielectrics. Phase self-modulation of the polarization components of travelling modes and power-dependent phase shifts between these modes are shown to act as the catalysts for deep amplitude and polarization modulation of trapped modes in a strip waveguide. The diversity of these effects, expanded due to geometric birefringency of waveguides close to cutoff and light-induced anisotropy, gives an effective tool for design of new non-linear fibre modulators, polarizers, filters, and sensors. In contrast to the well-known twin-core waveguide couplers, modulated single-core couplers for co- and counter-propagating modes are discussed.

Mirrors with tunable reflectivity are discussed in Chapter 2 from the viewpoint of light-beam switching and wave-pulse shaping. These mirrors based on electrically and thermally driven non-linear interfaces are shown often to operate in bistable regimes, that provide the jump-like transitions between the intensities, phases, and polarizations of incident and reflected waves. Thermomagnetic phenomena as well as corrugation and multi-layer coating of interfaces prove to be useful for deep modulation of reflected and transmitted wave beams.

The marathon race in the field of pulse narrowing and compression inspired by the prospects of ultra-short pulses used in optical computers and communication systems is considered in Chapter 3. Unlike the traditional approach reducing the rich variety of non-linear pulse processes to the properties of solitons in the framework of so-called 'solitonics', the advantages and shortcoming of wide classes of both soliton and non-soliton pulses in different TE and TM waveguide modes are treated in this chapter. Realistic models of non-linear propagation of short pulses are discussed, the power-dependent dissipation, third-order dispersion, and inertia of non-linearity being taken into account.

A detailed overview of each chapter is given in the relevant chapter's introduction; moreover, each chapter is completed by a short conclusion clarifying the limited validity of some assertions.

The mathematical background of this book is connected with the non-linear Schrödinger equation and systems of such equations. This name for the equations describing the dynamics of non-linear pulses in the dispersive media is widely used, although Schrödinger himself never dealt with these problems. An investigation of the influence of initial amplitude–phase structure of the localized field on the spatial scales and energetic parameters of its tuning is performed in the framework of these equations. Simplified analytical methods for solution of such problems are demonstrated in detail in this book. The systematic use of conservation laws is shown to be a master key in the understanding of dispersionless phenomena. Research on dispersive effects is based on analytical solutions obtained by a variational approach and non-linear geometrical optics. The results discussed are proved in a 'physical way' and mathematical calculations are restricted and standardized as much as possible.

Unlike the numerous monographs containing in-depth and all-round overviews of non-linear fibre optics (see for example, Agrawal 1989), this book points out some new perspectives in the field, that hitherto have been insufficiently elaborated. Hence some well-known effects, such as waveguide attenuation, asymptotic properties of solitons, or stimulated Raman and Brillouin scattering, are considered as far as these effects have an influence on the novel phenomena under discussion. In contrast attention is drawn here to concepts of energy-dependent periodicity, gyrotropy, reflection, mode conversion, and non-reciprocity, which are being utilized today. Using the picturesque comparison given by Professor Sh. L. Glashow in his Nobel Lecture in 1979, one may characterize the subject under discussion not as a 'tapestry' but only as 'the threads in a tapestry'.

This book may be of interest to different readers:

- *For scientists* interested in the basic physics this book may stimulate attempts to extend the knowledge of dynamics of non-linear wave distributed systems as a new field of physics. The applications of this field to optoelectronics design is expected to provide a key to optical technology and computational plenty. The use of waveguide circuits gives unique opportunities for modelling of both dynamical and stochastic processes in distributed systems including bifurcations and the trend to chaos. The remarkable similarity in non-linear behaviour of localized wave fields in dispersive media of different physical natures (fibres, crystals, plasma) promotes the development of intuition in solution of relevant problems in cross-disciplinary physics. Lastly, the beauty and uncommonness of non-linear Schrödinger equations stimulate the introduction of novel physical concepts and images to modern optoelectronics.

- *For designers and users* of non-linear integrated optics this book can provide information about the latest advances in a new generation of power-dependent all-optical tunable devices, including such key elements as switches, filters, polarizers, modulators, coherent sensors, tunable mirrors, mode sorters, power discriminators and limiters, single-channel mode couplers, and other elements for control of light by light.

- *For lecturers and students* the arrangement of information within each section is suitable for didactic goals and instructive analysis of several problems, highlighting the influence of physically meaningful parameters on the effects discussed. The detailed mathematical approach and numerous illustrations may be useful for self-study. The combined information presented here permits one to grasp, in particular, the main ideas of a new topic of applied physics provisionally named as 'Computers for optics and optics for computers'.

During the last decade many people have contributed to the author's understanding of this field either knowingly in scientific collaboration and discussions or unwittingly through moral support. Among those who have contributed scientifically I thank my colleagues Drs I. Kolchanov, A. Shepelev, and M. Zuev. I appreciate highly discussions with Professors T. Arecchi, F. Bunkin, L. Keldysh, F. Kneubuhl, A. Schmidt, A. Snyder, R. Stalio, B. Wilhelmi, and E. Wintner. It is my pleasure to thank Professors A. Kaplan, T. Wilson, and V. Zakharov for their invisible influence on this work. Once again I thank my family for patience and encouragement.

In considering the intended readership for this book I have divided them conventionally into the three above-mentioned groups. I hope to avoid a reduction to the groups described by a playful poem written at one laser conference:

> Some people, who are 'crazy',
> Spend their lives studying lasers;
> Most people, who are lazy,
> Sleep on books about lasers.

Moscow
September 1992

A. B. S.

Contents

1.

Fundamentals of intensity-dependent phase phenomena for CW operations in waveguide circuits

1.1 Introduction: basic non-linear phase effects in bulk isotropic media

The theoretical analysis of non-linear integrated and waveguide optics is based on the solution of Maxwell's equations for spatially localized fields. Realization of series of non-linear index-related effects is implemented by high intensity of the wave fields, achieved due to the localization of wave energy flow in guided systems. However, we may discuss the essential physics of some of these phenomena using the simple model of powerful plane waves in an unbounded homogeneous non-linear medium. Thus, the basic non-linear phase effects are described by means of this model. The task of this section consists in evaluation of energetic parameters and spatial scales of intensity-dependent phase phenomena within the scope of such a simplified approach.

The set of Maxwell's equations for a dielectric isotropic material can be written in the form

$$\operatorname{rot} \mathbf{E} = -\mu_0 \frac{\partial \mathbf{H}}{\partial t}, \tag{1.1}$$

$$\operatorname{rot} \mathbf{H} = \frac{\partial \mathbf{D}}{\partial t}, \tag{1.2}$$

$$\operatorname{div} \mathbf{D} = 0, \tag{1.3}$$

$$\operatorname{div} \mathbf{H} = 0. \tag{1.4}$$

$\mathbf{H}$ and $\mathbf{E}$ are the magnetic and electric field vectors, t is the time variable, ε_0 and μ_0 are the dielectric permittivity and magnetic permeability of vacuum. We assume that the magnetic properties of the material are those of vacuum. The electric displacement $\mathbf{D}$ can be expressed as

$$\mathbf{D} = \varepsilon_0 n_0^2 \mathbf{E} + \mathbf{D}_N. \tag{1.5}$$

Here n_0 is the index of refraction at low intensity, connected with the relative dielectric permittivity of the material $\varepsilon = n_0^2$. $\mathbf{D}_N$ is the non-linear component of electric displacement. Since materials with third-order non-linearity due to the Kerr effect are widely utilized in integrated and waveguide optics, these materials will be discussed below. Maker *et al.* (1964) showed that in the general case of an arbitrary type of third-order non-linear susceptibility the complex non-linear component of electric displacement $\mathbf{D}_N$ may be written as

$$\mathbf{D}_N = \varepsilon_0 \chi [A|\mathbf{E}|^2\mathbf{E} + B(\mathbf{E})^2\mathbf{E}^*] \tag{1.6}$$

where χ is a constant of non-linear interaction for a single linearly polarized plane wave, and A and B are dimensionless constants characteristic of the particular mechanism of non-linearity. In a lossless medium A and B are real quantities and

$$A + B = 1. \tag{1.7}$$

The value of ratio BA^{-1} depends on the properties of the material. Thus, $BA^{-1} = 3$ in liquids, and $BA^{-1} = 0.5$ in crystals; for electrostriction $B = 0$.

In dealing with non-magnetized media, it is worthwhile to derive from the system (1.1–1.3) a single equation governing the behaviour of the electric field $\mathbf{E}$. By combining the two equations (1.1) and (1.2) we obtain

$$\triangle \mathbf{E} - \operatorname{grad} \operatorname{div} \mathbf{E} - \mu_0 \frac{\partial^2 \mathbf{D}}{\partial t^2} = 0. \tag{1.8}$$

Here $\triangle$ is the Laplace operator. The condition

$$\operatorname{div} \mathbf{E} = 0, \tag{1.9}$$

following from eqn (1.3) for a homogeneous unbounded medium, results in vanishing of the term $\operatorname{grad} \operatorname{div} \mathbf{E}$ in eqn (1.8), although this term will be shown in Section 1.4 to become important for the description of guided systems, especially in the vicinity of cutoff frequencies.

We consider here only strictly time-harmonic fields whose time dependence, in complex notation, can be expressed as $\exp(-i\omega t)$. The radian frequency ω is related to the actual frequency ν by

$$\omega = 2\pi\nu. \tag{1.10}$$

The plane wave travelling in the z-direction is described by the function

$$\exp[i(\beta z - \omega t)]. \tag{1.11}$$

Here β is the wavenumber

$$\beta = \omega_0 n_0 \sqrt{\mu_0 \varepsilon_0} \tag{1.12}$$

The 'vacuum' parameters μ_0 and ε_0 are connected with the light velocity in free space, c, by the relation

$$c = (\mu_0\varepsilon_0)^{-1/2} \tag{1.13}$$

Substitution of function (1.12) gives the field eqn (1.8) describing the spatial evolution of electric field $\mathbf{E}$:

$$\beta^2 \mathbf{E} - \omega^2\sqrt{\mu_0\varepsilon_0}\, n_0^2 \mathbf{E} - \omega^2\sqrt{\mu_0\varepsilon_0}\, \chi[A|\mathbf{E}|^2\mathbf{E} + B(\mathbf{E})^2\mathbf{E}^*] = 0. \tag{1.14}$$

When the wave intensity is low and non-linearity tends to zero, eqn (1.14) is reduced to the well-known 'linear' dispersive equation:

$$\beta^2 = \frac{\omega^2 n_0^2}{c^2} \tag{1.15}$$

In order to simplify the notation we use in (1.15) the light velocity c from (1.13).

The non-linear component of electric displacement $\mathbf{D}_N$ (1.6) depends upon the constant χ. Typical values of this constant for the series of materials utilized in integrated and waveguide optics are small; thus, $\chi = (1 \sim 3) \times 10^{-13}$ esu for silica fibres. The non-linear component of electric displacement $\mathbf{D}_N$ is usually much smaller than the linear component in the fibres. The influence of this non-linear perturbation of dielectric permittivity on a wave's field variations may be supposed to accumulate slowly in the course of wave propagation. Therefore we assume that the complex amplitude of travelling wave

$$\mathbf{E} = \mathbf{E}_0 e^{i(\beta z - \omega t)} \tag{1.16}$$

may be described as a product of a slowly varying factor $\mathbf{E} = \mathbf{E}(z)$ and a rapidly varying exponential factor. Let us substitute eqn (1.16) into eqn (1.8) and calculate the second derivative $\partial^2 E/\partial z^2$, taking into account the pressence of two different spatial scales of variation of the field. Making use of eqn (1.14), one can obtain the reduced equation governing the non-linear evolution of the wave's amplitude:

$$2i\beta\frac{\partial \mathbf{E}}{\partial z} + \frac{\omega^2}{c^2}\chi[A|\mathbf{E}|^2\mathbf{E} + B(\mathbf{E})^2\mathbf{E}^*] = 0. \tag{1.17}$$

Equation (1.17) permits us to analyse some basic non-linear phase effects existing in both unbounded and guided media, although the energetic parameters and spatial scales of similar processes in these cases may distinguish considerably. Firstly, we provide an insight into these processes in unbounded media by discussing the solutions of eqn (1.17) for different polarizations of the wave field $\vec{\mathbf{E}}$.

Simultaneously, we shall demonstrate the standard way of analysis of the problems discussed.

Some major tendencies in development of non-linear phase-dependent processes in bulk media are outlined below:

A. Phase self-modulation; this effect is helpful for the series of devices using power-dependent interference of waves.
B. Mutual phase modulation of a pair of counter-propagating fields; the important property of this phenomenon is the non-reciprocity of phase shifts obtained by interacting fields. Such phase non-reciprocity is perspective for the flexible tuning of radiation modes overlapping in all-optical circuits.
C. Self-action of elliptically polarized wave. This interaction of polarization components provides three useful effects:

1. Light induced gyroptropy of initially isotropic medium;
2. Non-linear spatial oscillations of polarization structure of the field;
3. Tuning of depth and period of coupled spatial cross-modulation of polarization components controlled by initial amplitude-phase characteristics of the input wave, it's power being kept constant. This concept of self-action of elliptically polarized field may be considered as a simple model of non-linear evolution of higher TE modes in a real waveguide.

These gradually complicating series of problems will be shown in this chapter to have a close relationship to non-linear phase effects in waveguide circuits. Thus, it is expedient to preface the overview of such effects, given at the end of this section, with some standard solutions of the above-mentioned problems A, B, and C.

A. Intensity-dependent phase shift of a linearly polarized wave   Let us seek the solution of eqn (1.17) in a form

$$E = E_0 f(z) e^{i\varphi(z)} \tag{1.18}$$

Substitution of eqn (1.18) into eqn (1.17) and separating real and imaginary parts of eqn (1.17) permits us to write two equations for the dimensionless amplitude $f(z)$ and phase $\varphi(z)$ respectively:

$$\frac{\partial f}{\partial z} = 0, \qquad \frac{\partial \varphi}{\partial z} = \frac{\omega^2}{2\beta c^2} \chi |E_0|^2 \tag{1.19}$$

The function f and φ must satisfy initial conditions at the point $z = 0$:

$$f\big|_{z=0} = 1, \qquad \varphi\big|_{z=0} = 0. \tag{1.20}$$

The solution of the simple system (1.19–1.20) is obvious:

$$E_0 = \text{const}; \qquad \varphi = \frac{\pi}{2} \frac{z}{L_n} \operatorname{sign} \chi. \tag{1.21}$$

The appearance of the discontinuous function sign χ in eqn (1.21) is connected with the properties of materials, being in the range of positive values of parameter χ for so-called 'focussing media' and in the range of negative χ for 'defocusing media'

$$L_n = \frac{\lambda_0 n_0}{2|\chi|E|^2|}. \tag{1.22}$$

Thus non-linearity results in the formation of an additional phase shift, while the amplitude remains constant. The phase perturbation is characterized by 'non-linear length' L_n (1.22), related to the phase shift and equal to $\pi/2 \operatorname{sign} \chi$.

B. Light-induced non-reciprocity of counter-propagating waves The non-linear index-related perturbation of a Kerr-like medium in the field of two linearly polarized counter-propagating waves may be analysed in the framework of eqn (1.17). Let us consider for simplicity waves of the same frequency ω and amplitudes E_1 and E_2, the polarization vectors $\mathbf{E}_1$ and $\mathbf{E}_2$ being parallel. The spatial dependence of amplitude of total field may be written as

$$\mathbf{E} = \mathbf{E}_1 e^{i\beta z} + \mathbf{E}_2 e^{-i\beta z} \tag{1.23}$$

The non-linearity of the medium produced by the field (1.23), may be analysed by means of substitution of the total field (1.23) into eqn (1.6) and use of relation (1.7). It is essential that the non-linear components of electric displacement $\mathbf{D}_N$ are different for opposite directions of wave propagation:

$$\mathbf{D}_{N1} = \varepsilon_0 \chi (|\mathbf{E}_1|^2 + 2|\mathbf{E}_2|^2) \mathbf{E}_1 e^{i\beta z},$$
$$\mathbf{D}_{N2} = \varepsilon_0 \chi (|\mathbf{E}_2|^2 + 2|\mathbf{E}_1|^2) \mathbf{E}_2 e^{-i\beta z}. \tag{1.24}$$

The rapidly varying terms, proportional to $e^{\pm 3i\beta z}$, are omitted here.

Thus, induced non-reciprocity is shown to be produced in the system of counter-propagating plane waves. The non-reciprocity is caused by differences in optical path lengths, varying as functions of the light intensity, the geometric path lengths of the counter-propagating waves being equal. The non-reciprocal factors in eqn (1.24) result from the formation of a non-linear index grating that leads to the cross interaction of $\mathbf{E}_1$ and $\mathbf{E}_2$. This grating is caused by the periodic term in the expression for non-linear displacement $\mathbf{D}$.

$$\mathbf{E}_1 \mathbf{E}_2^* e^{2i\beta z} + \text{c.c} = 2|E_1 E_2| \cos(2\beta z + \varphi).$$

Here this cross interaction does not lead to mutual back-reflection the counter-propagating waves.

Substitution of expressions (1.24) into the non-linear wave equation (1.17) yields that the amplitudes of interacting waves remain constant, while their phase shifts are different:

$$\varphi_1 = \frac{\omega \chi z}{2 c n_0} \left(|E_1|^2 + 2|E_2|^2 \right),$$

$$\varphi_2 = \frac{\omega \chi z}{2 c n_0} \left(|E_2|^2 + 2|E_1|^2 \right). \tag{1.25}$$

This example of non-reciprocity relates to the trivial geometry of interacting wave fields, when both waves are polarized linearly with the polarization vectors being parallel. The effects of non-reciprocity produced by other polarization structures of counter-propagating waves may be analysed by means of generalized equations (Kaplan 1983):

$$2i\frac{\partial \mathbf{E}_1}{\partial z} + \frac{\omega \varepsilon_0 \chi}{c n_0} \left\{ A \left[\mathbf{E}_1 \left(|\mathbf{E}_1|^2 + |\mathbf{E}_2|^2 \right) + \mathbf{E}_2 (\mathbf{E}_1 \mathbf{E}_2^*) \right] \right.$$

$$\left. + B \left[2\mathbf{E}_2^* (\mathbf{E}_1 \mathbf{E}_2) + \mathbf{E}_1^* (\mathbf{E}_1)^2 \right] \right\} = 0,$$

$$2i\frac{\partial \mathbf{E}_2}{\partial z} + \frac{\omega \varepsilon_0 \chi}{c n_0} \left\{ A \left[\mathbf{E}_2 \left(|\mathbf{E}_1|^2 + |\mathbf{E}_2|^2 \right) + \mathbf{E}_1 (\mathbf{E}_2 \mathbf{E}_1^*) \right] \right.$$

$$\left. + B \left[2\mathbf{E}_1^* (\mathbf{E}_1 \mathbf{E}_2) + \mathbf{E}_2^* (\mathbf{E}_2)^2 \right] \right\} = 0. \tag{1.26}$$

The system (1.26) describes a separate conservation of energy flow in each of interacting waves. Making use of eqn (1.26) one can see that the partial non-linear susceptibilities $\delta \varepsilon_j$, related to the respective waves $\mathbf{E}_j$, are

$$\Delta \varepsilon_j = \varepsilon_0 \chi \left[\alpha |\mathbf{E}_j|^2 + (\alpha + \beta) |\mathbf{E}_{3-j}|^2 \right].$$

Here the aforesaid case of linearly polarized waves ($\mathbf{E}_1 \| \mathbf{E}_2$) is described by values $\alpha = \beta = 1$. In contrast, $\alpha = 1$, $\beta = -B$ if the waves are linearly polarized with $\mathbf{E}_1 \perp \mathbf{E}_2$. Finally, co- and counter-rotating circularly polarized waves are characterized by the values $\alpha = \beta = 1 - B$ and $\alpha = 1 - B$, $\beta = 2B$, respectively.

Alongside non-reciprocity, induced by a powerful wave field, the fundamentals of light-induced gyrotropy may also be described by a model of plane waves in a bulk isotropic medium.

C. Light-induced gyrotropy of one-directional waves In order to consider the self-action of elliptically polarized waves we introduce

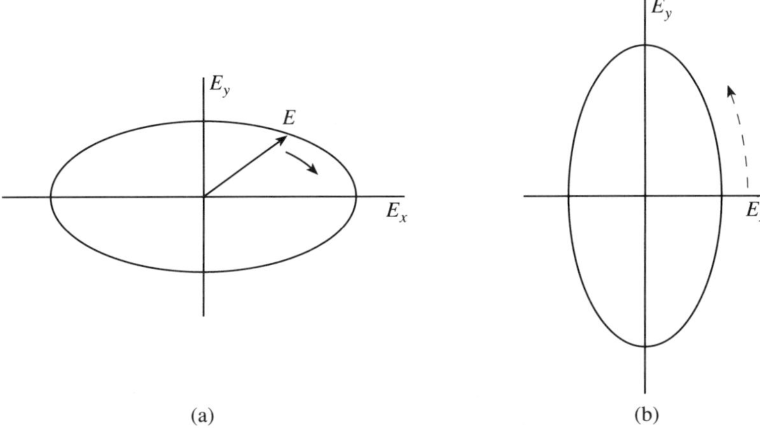

Fig. 1.1 Intensity-dependent rotation of the polarization ellipse. (a) Polarization state in linear theory; the arrow shows the clockwise direction of rotation of the electric field vector **E** over the ellipse, $\varphi_0 = \pi/2$. (b) The dotted arrow indicates the non-linear turning of the ellipse in the counter-clockwise direction.

the Cartesian coordinate system (x, y) in the plane $z = 0$, the axes X and Y coinciding with the directions of the major and minor axes of the ellipse of polarization, respectively (Fig. 1.1). This problem may be treated by analogy with the previous one, although the analysis is complicated by presence of two orthogonal polarization components E_x and E_y.

It is convenient to represent each of the complex polarization components E_x and E_y as the product of amplitude and phase factors:

$$E_x = ae^{i\varphi_1}; \qquad E_y = be^{i\varphi_2}. \tag{1.27}$$

Substitution of these relations into the master equation (1.17) and separating real and imaginary parts yields the following system:

$$2\beta \frac{\partial \varphi_1}{\partial z} - \frac{\omega^2 \chi}{c^2} \left[A(a^2 + b^2) + B(a^2 + b^2 \cos 2\varphi) \right] = 0, \tag{1.28}$$

$$2\beta \frac{\partial \varphi_2}{\partial z} - \frac{\omega^2 \chi}{c^2} \left[A(a^2 + b^2) + B(a^2 \cos 2\varphi + b^2) \right] = 0, \tag{1.29}$$

$$2\beta \frac{\partial a}{\partial z} + \frac{\omega^2 \chi B}{c^2} ab^2 \sin 2\varphi = 0, \tag{1.30}$$

$$2\beta \frac{\partial b}{\partial z} - \frac{\omega^2 \chi B}{c^2} ba^2 \sin 2\varphi = 0. \tag{1.31}$$

Here the new function φ, describing the phase shift φ between E_x and E_y components,

$$\varphi = \varphi_2 - \varphi_1 \tag{1.32}$$

is introduced. The unknown functions a, b, and φ must satisfy initial conditions at the point $z = 0$:

$$a\big|_{z=0} = a_0; \qquad b\big|_{z=0} = b_0; \qquad \varphi\big|_{z=0} = \varphi_0. \tag{1.33}$$

The analysis of the system (1.28–1.32) may be simplified by use of dimensionless functions f_1, f_2 and dimensionless variable z:

$$a = a_0 f_1(z); \qquad b = b_0 f_2(z); \tag{1.34}$$

$$\tau = \frac{z}{L_0}; \qquad L_0 = \frac{\lambda_0 n_0}{B \chi a_0^2 \pi}. \tag{1.35}$$

Substracting eqn (1.28) from eqn (1.25) we obtain an equation governing the function φ (1.32):

$$\frac{\partial \varphi}{\partial \tau} = -2 \sin^2 \varphi (f_1^2 - f_2^2). \tag{1.36}$$

It is convenient to introduce parameter μ describing the ratio of axes of polarization ellipse:

$$\mu = \frac{b_0}{a_0} \leqslant 1 \tag{1.37}$$

The pair of equations (1.30), (1.31) is reduced by the notation (1.34), (1.35) to dimensionless form:

$$\frac{\partial f_1}{\partial \tau} = -f_2^2 f_1 \sin 2\varphi, \tag{1.38}$$

$$\frac{\partial f_2}{\partial \tau} = f_1^2 f_2 \sin 2\varphi. \tag{1.39}$$

Initial conditions (1.33) to the set of equations (1.36, 1.38), (1.39) may be written as

$$f_1\big|_{\tau=0} = 1; \quad f_2\big|_{\tau=0} = \mu; \qquad \varphi\big|_{\tau=0} = \varphi_0. \tag{1.40}$$

Thus, we derived a self-consistent system of equations describing the non-linear amplitude–phase evolution of elliptically polarized waves.

The system (1.36, 1.38), (1.39) has two invariants:

$$f_1^2 + f_2^2 = 1 + \mu^2, \tag{1.41}$$

$$f_1 f_2 \sin \varphi = \mu \sin \varphi_0. \tag{1.42}$$

Invariant (1.41) describes the conservation of an energy flow, whereas invariant (1.42) relates to the conservation of a momentum flux of the field.

Before proceeding further it is useful to obtain two auxiliary relations, following from the master system (1.38, 1.39):

$$\frac{\partial (f_1 f_2)}{\partial \tau} = f_1 f_2 (f_1^2 - f_2^2) \sin 2\varphi, \tag{1.43}$$

$$\frac{\partial (f_1^2 - f_2^2)}{\partial \tau} = -4 f_1^2 f_2^2 \sin 2\varphi. \tag{1.44}$$

The right-hand side of eqn (1.44) may be transformed by means of invariant (1.48). Supposing for simplicity

$$\varphi_0 = \frac{\pi}{2}; \qquad f_1 f_2 \sin \varphi = \mu, \tag{1.45}$$

we obtain from (1.44)

$$\frac{\partial (f_1^2 - f_2^2)}{\partial \tau} = -8 f_1 f_2 \cos \varphi. \tag{1.46}$$

Now let us differentiate eqn (1.46) with respect to variable τ. Replacement of the derivatives $\partial / \partial \tau \, (f_1 f_2)$ and $\partial \varphi / \partial \tau$ by means of eqn (1.43) and (1.36), respectively, leads to a simple equation governing the function $f_1^2 - f_2^2$:

$$\frac{\partial^2 (f_1^2 - f_2^2)}{\partial \tau^2} + 16 (f_1^2 - f_2^2) = 0. \tag{1.47}$$

The solution of this equation, satisfying initial conditions

$$f_1^2 - f_2^2 \big|_{\tau = 0} = 1 - \mu^2 \tag{1.48}$$

may be written as

$$f_1^2 - f_2^2 = (1 - \mu^2) \cos 4\mu z. \tag{1.49}$$

Combining the solution (1.49) with the invariant (1.41) we finally obtain, using trigonometric identities

$$f_1^2 = \cos^2 2\mu\tau + \mu^2 \sin^2 2\mu\tau,$$
$$f_2^2 = \sin^2 2\mu\tau + \mu^2 \cos^2 2\mu\tau. \tag{1.50}$$

Thus, the non-linear interaction of components of the polarization ellipse results in periodic oscillations of the amplitudes of components $E_x = af_1$ and $E_y = a\mu f_2$. The principal axes of this ellipse $E_{m,s}$ at each point may be calculated by means of the well-known geometric expression

$$E_{m,s} = \tfrac{1}{2}[\sqrt{(|E_x|^2 + |E_y|^2 + 2|E_x E_y| \sin \varphi)^{1/2}}$$
$$\pm \sqrt{(|E_x|^2 + |E_y|^2 + 2|E_x E_y| \sin \varphi)^{1/2}}] \qquad (1.51)$$

If we substitute the expressions (1.27) and (1.34) into (1.51) and utilize invariants (1.41) and (1.42), we obtain that the principal half-axes of polarization ellipse remain constant in the course of evolution. Here the evolution may be considered as slow rotation of the polarization ellipse in a direction opposite to the direction of rapid rotation of the electric vector **E** providing the elliptical polarization. The angle θ between the initial direction of major axis and its direction at some arbitrary point τ may be calculated from the equation

$$\cos^2 \theta + \mu^2 \sin^2 \theta = f_1^2. \qquad (1.52)$$

Using the solution (1.50) we obtain that the major axis rotates monotonically:

$$\theta = 2\mu\tau. \qquad (1.53)$$

The half-period of this rotation, τ_0, related to the turning angle $\theta = \pi/2$, may be expressed using (1.50) as

$$\tau_0 = \frac{\pi}{4\mu}; \qquad z_{\pi/2} = \frac{\lambda_0 n_0}{4B\mu\chi a^2}. \qquad (1.54)$$

Finally, the phase shift φ oscillates in the course of propagation according to the expression following from invariant (1.42) for $\sin \varphi_0 = 1$:

$$\sin^2 \varphi = \frac{\mu^2}{f_1^2 f_2^2} = \frac{1}{1 + [(1 - \mu^2)^2/4\mu^2] \sin^2 4\mu\tau} \qquad (1.55)$$

Thus the phase shift varies between the values $\varphi_{1,2}$ (Fig. 1.2) determined from the equations

$$\sin \varphi_{1,2} = \frac{2\mu}{1 + \mu^2}. \qquad (1.56)$$

The period of phase oscillation $\tau_\varphi = \pi(4\mu)^{-1}$ proves to be half the period of ellipse rotation (1.54). It is essential that light-induced gyrotropy is produced, unlike the light-induced non-reciprocity, by a non-zero value of the parameter B (1.54).

We conclude this section with two remarks:

1. It is worthwhile emphasizing the useful relation between the non-linear perturbations of dielectric permittivity $\Delta\varepsilon$ and refractive index Δn,

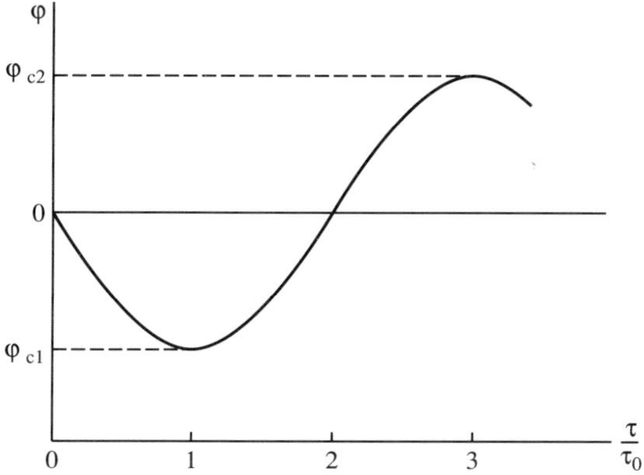

Fig. 1.2 Non-linear oscillations of phase shift φ between E_y and E_x components of the polarization ellipse ($\mu = 0.5$).

$$\Delta\varepsilon = 2n_0\,\Delta n. \tag{1.57}$$

Thus, the substitution of eqn (1.57) into eqn (1.22) transforms the definition of non-linear length L_n, related to formation of additional phase shift of $\pi/2$, to a physically meaningful form:

$$L_n = \frac{\lambda_0}{4\,|\Delta n|} \tag{1.58}$$

The distance between two points in free space with shift $\varphi = \pi/2$ between the wave field oscillations at these points is known to be equal to $\lambda_0/4$. The non-linear perturbation $|\Delta n|$ in the denominator of eqn (1.58) enlarges this distance $|\Delta n|^{-1}$ times.

2. For quick evaluation of non-linear perturbation of refractive index Δn in Kerr-like media it is convenient to use the following form of the Kerr law:

$$\Delta n = n_2 P \tag{1.59}$$

Here P labels the density of power flow, and the parameter n_2 for glasses is around $n_2 = 3 \times 10^{-16}\,\mathrm{cm^2\,W^{-1}}$.

This analysis illustrates some fundamentals of the physics of non-linear processes characterizing the controlled evolution of phases and polarization structure of plane waves in Kerr-like lossless dielectric. Starting with examples of these phenomena in bulk media it should be comparatively straightforward to study the relevant effects in planar and strip waveguides and optical fibres. Manifestations of the same tendencies in guiding systems may be diversified considerably due to mode phenomena. To develop the insight of power-dependent tuning of modes and it's using for design of all-optical devices this chapter is reorganized as follows.

Since the non-linear dynamics of trapped radiation will be shown to depend upon its modal purity and amplitude-phase distribution at the waveguide input, the control of these characteristics is a vital factor for all-optical operations. The governed excitement of individual modes in a fibre by means of spatially unmodulated and modulated pumping beams is discussed in Section 1.2.

The influence of polarization structure of excited individual modes on their phase self-action is analysed in Section 1.3. Unlike the usual phase self-modulation of linearly polarized TE_{01} mode generalizing the well known bulk non-linear effect (Section 1.1, part A), the difference of phase shifts for each polarization component of TE_{11} or TM_{11} modes results in light induced gyrotropy. This power-dependent gyrotropy is characterized by geometric effects; thus, the dependence of the above-mentioned phase shifts in a strip waveguide upon the ratio of its sides is outlined.

The enhancement of such waveguide geometry effects in the non-linear dynamics of excited modes closed to cutoff is accentuated in Section 1.4. The self-action of higher TE and TM modes excites the spatial modulation of its amplitude-phase structure. The deepening of modulation and shortening of its spatial period near by the cutoff of each mode are illustrated.

Unlike the above-mentioned self-action of individual modes, the governed interaction of coupled modes is considered in Section 1.5, which is devoted to non-linear directional couplers and phase sensors. The power transfer between two adjacent guiding channels (twin-core coupler) and between two modes in one slightly anisotropic channel (single-core coupler) is investigated, the initial phase shift between modes, their power, and linear coupling characteristics being taken into account. The spatial oscillations of coupled modes depending upon the phase shift between them and sensor-related applications of this effect utilizing non-linear birefringent waveguides are emphasized.

The continuation of this analysis providing the elaboration of more sophisticated single-core couplers and logical elements is given in Section 1.6. The examples of resonant interaction of co-propagating fields in one channel, such as phase-dependent and phase-independent types of conversion of two and three modes respectively, are also demonstrated.

An opposite case, pertaining to interaction of conterpropagating modes in a single-core channel is the topic of Section 1.7. The power transfer between these modes in the spatially modulated guiding system (e.g. Bragg filter) is responsible for intensity-depended tunable transmittivity of this system. The amplification of non-reciprocal phase modulation of interacting modes close to waveguide cutoff is noted.

The enhancement of interaction of above-mentioned counter-propagating modes due to resonant properties of oprical circuits is discussed in Section 1.8. Use of these properties in switching and sensing optical schemes is shown.

1.2 Controlled excitement of modes in the waveguide

Propagation of wave fields in guided systems stimulates the amplification of non-linear processes owing to considerable concentration of energy flow. Moreover, energetic parameters and spatial scales of such processes in multimode guide systems depend upon the mode of propagation. Thus, in order to govern the evolution of the trapped radiation it is necessary to provide controlled excitation of individual modes in the waveguide. This problem appears in the elaboration of sensors, holography devices and distant communication channels, based on multimode fibres. Since we shall discuss below the merits and short-comings of some selected modes from the standpoint of non-linear processes, it is worthwhile illustrating briefly here the possibilities of controlled excitement of these modes by laser beams incident on the waveguide's input.

The mode structure of wave travelling in the z-direction is formed by combination of waveguide's eigenmodes:

$$E_j = E_{0j} \sum_{n=0}^{m} \sum_{p=1}^{l} \xi_{npj} \, \psi_{np} \, e^{i(\beta_{np} z - \omega t)} \tag{1.60}$$

Here E_{0j} is the jth polarization component of trapped radiation; ψ_{np} are the eigenfunctions of the wave equation, describing the transverse structure of guided modes; coefficients ξ_{npj} determine the contribution of the npth mode to the jth component of the total wave field. The propagation constant β_{np} is connected with transverse component of wave vector $k_\perp$:

$$\beta_{np} = \sqrt{\frac{\omega^2 n_0^2}{c^2} - k_{\perp np}^2} \, . \tag{1.61}$$

n_0 is the value of the refractive index, related to low intensities. The spectrum of quantized values $k_{\perp np}^2$ is determined by the profile of refractive index, $n(\rho)$.

It is worth noting the circularly-shaped waveguide with a parabolic profile of refractive index and radius a of waveguide core:

$$n(\rho) = n_0 \left(1 - \frac{\Delta \rho^2}{a^2} \right); \quad 0 < \Delta \ll 1; \quad \rho \leq a. \tag{1.62}$$

In order to demonstrate some trends in controlled excitement of Gaussian modes in such waveguides, let us restrict ourselves for definitness to

TE modes. Substitution of profile (1.62) into the trivial wave equation and use of dimensionless variables yields the equation governing the longitudinal component of magnetic vector B_z of the TE mode:

$$\frac{\partial^2 B_z}{\partial u^2} + \frac{1}{u}\frac{\partial B_z}{\partial u} + \frac{1}{u^2}\frac{\partial^2 B_z}{\partial \varphi^2} + [q^2 - u^2]B_z = 0, \qquad (1.63)$$

$$u = \frac{\rho}{\rho_0}; \qquad q^2 = k_\perp^2 \rho_0^2; \qquad \rho_0 = \left(\frac{ca}{\omega n_0 \sqrt{2\Delta}}\right)^{1/2}. \qquad (1.64)$$

The dimensionless radius u and azimuth variable φ characterize the point (u, φ) in the waveguide's cross-section.

The eigenfunctions of eqn (1.63) are known to be Gaussian–Laguerre modes. The transverse components of the electric field of such TE modes are expressed by means of longitudinal magnetic components B_z:

$$E_\rho = \frac{i\omega}{ck_\perp \rho_0}\frac{1}{u}\frac{\partial B_z}{\partial \varphi}; \qquad E_\varphi = \frac{-i\omega}{ck_\perp \rho_0}\frac{\partial B_z}{\partial u}.$$

In view of forthcoming applications it is convenient to write here the field distributions for some axisymmetrica TE_{0p} Gausse–Laguerre modes. Amplitude E_{0j} (1.60) relates in this geometry to the E_φ component described by equation

$$\frac{\partial^2 E_\varphi}{\partial u^2} + \frac{1}{u}\frac{\partial E_\varphi}{\partial u} - \frac{E_\varphi}{u^2} + (q^2 - u^2)E_\varphi = 0. \qquad (1.65)$$

Eigenfunctions ψ_{0p} (1.2.1) satisfying the normalization condition

$$\int_0^\infty |\psi_{0p}(u)|^2 u\,du = 1 \qquad (1.66)$$

may be written as

$$\psi_{01} = C_1 u e^{-u^2/2}; \qquad \psi_{02} = c_2 u\left(1 - \frac{u^2}{2}\right)e^{-u^2/2},$$

$$\psi_{03} = C_3 u\left(1 - u^2 + \frac{u^4}{6}\right)e^{-u^2/2}; \qquad C_p = \sqrt{2p} \qquad (1.67)$$

The radial distributions ψ_{01} and ψ_{02}, useful for forthcoming calculations of non-linear regimes, are shown in Fig. 1.3. The characteristic radial scale ρ_0 is given in eqn (1.64); the spectrum q_p^2 is quantized:

$$k_\perp^2 = \frac{4p}{\rho_0^2}, \qquad p = 1, 2, 3. \qquad (1.68)$$

The interest in Gauss–Laguerre modes is twofold. Firstly, these localized fields are known to possess properties important for their

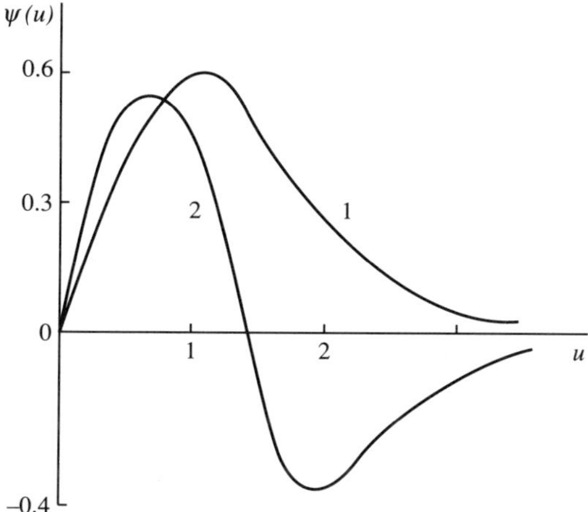

Fig. 1.3 Dimensionless radial distributions of electric amplitudes $e(u)$ in the TE_{01} (curve 1) and TE_{02} (curve 2) modes in a circularly-shaped waveguide with parabolic profile of refractive index (eqn (1.62)).

utilization in systems for controlled spatial modulation of radiation beams. On the other hand, use of these modes often provides a successful mathematical approach to solution of non-linear waveguide problems in analytical form, avoiding computer simulation. However, the conditions for excitement of these modes are determined mainly by linear input process. Since such processes are considered the subject of a spatial part of linear fibre optics (Snyder and Love 1983), we will mention below only some results that are useful for evaluation of power of radiation trapped in a waveguide. We shall illustrate the possibilities of controlled excitement of these modes both by pumping beam directly and by pumping beam passed through a spatial modulator.

1.2.1 *Direct excitement of Gauss–Laguerre modes by Gaussian pumping beams*

The properties of series of elements of integrated and waveguide optics, such as optical cavities, lens-like media and graduated optical waveguides, result in formation of Gaussian amplitude–phase distributions of light waves. Such distributions are formed as the eigenmodes of both open laser cavities with spherical mirrors and parabolic-index waveguides. Moreover, the Gaussian beams are known to conserve their spatial structure both in the course of propagation in free space and during passage through optical lenses. This circumstance suggests the use of Gaussian beams irradiated by lasers and focused by lenses for excitement of Gauss–Laguerre modes in parabolic-index fibres.

Let us consider the simple case of a paraxial Gaussian wave beam, restricted at the plane $z = 0$ by a circularly shaped aperture with radius ρ_{10}. Supposing the wave front at the initial cross-section of the beam $z = 0$ to be plane, we may obtain the complex wave amplitude of Gauss–Laguerre axisymmetric mode in a free space:

$$\psi_0 = \exp\left[-\frac{\rho^2}{\rho_1^2} + ik\left(z + \frac{\rho^2}{2R}\right) - i\,\text{arctg}\,\frac{z}{b}\right]. \qquad (1.69)$$

Here parameters ρ_1 and R, denoting the effective radius of the beam and the radius of curvature of its wave front, respectively, are expressed via the diffractive length b:

$$\rho_1 = \rho_{10}\sqrt{\left(1 + \frac{z^2}{b^2}\right)}; \qquad R = z + \frac{b}{z}; \qquad b = \frac{k\rho_{10}^2}{2}. \qquad (1.70)$$

Passage of Gaussian beam through the focusing lens leads to formation of Gaussian beam too. When this beam is incident on the parabolic-index fibre input, the efficiency of excitation of Gauss–Laguerre modes depends upon the geometric conditions of incidence. These conditions are determined by angle θ between beam's and the fibre's axes, the displacement $\Delta\rho$ of beam's centre from fibre's axis in the input plane of the fibre, and the ratio of effective radii of the pumping beam, ρ_1, and of the trapped mode, ρ_0. It is convenient to introduce dimensionless parameters characterizing these distortions:

$$\sigma = \left(\frac{\rho_1}{\rho_0}\right)^2; \qquad p_0 = 2\left(\frac{\Delta\rho}{\rho_0}\right); \qquad s^2 = \left(\frac{\rho_0\sin\theta}{a\sin\theta_\text{m}}\right)^2. \qquad (1.71)$$

Here the value ρ_0 is given in (1.64) and θ_m is the waveguide's aperture. Omitting the cumbersome calculations, one may evaluate the influence of these distortions on the dimensionless intensity W of fundamental Gauss–Laguerre mode excited in parabolic-index waveguide using the result obtained by Grau *et al.* (1980):

$$W = \frac{4\sigma}{(1 + \sigma)^2}\exp\left(-\frac{p_0^2 + \sigma s^2}{1 + \sigma}\right). \qquad (1.72)$$

Decrease of distortions ($\sigma \Rightarrow 1, p_0 \Rightarrow 0, s \Rightarrow 0$) provides the growth of intensity up to its maximum value $W = 1$. Dependences (1.72), illustrated in Fig. 1.4–1.6, show the considerable decrease of trapped power due to distortions of pumping conditions (1.71). Side by side with the slackening of fundamental mode, the parasitic modes may be excited too. The modal purity of trapped radiation can be improved by means of special input mode filters, which are discussed below.

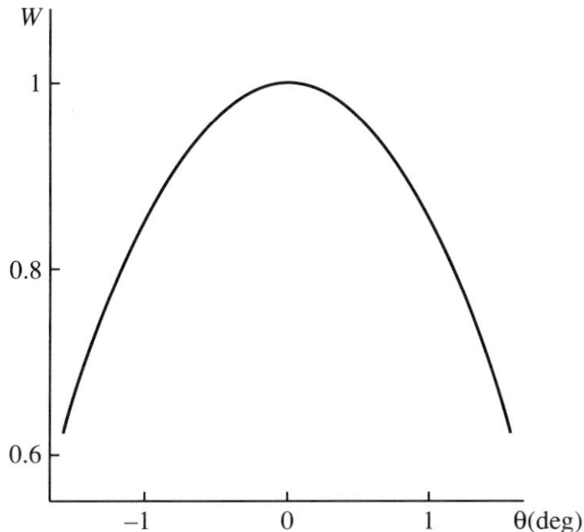

Fig. 1.4 Decrease of dimensionless power W of a guided TE_{01} mode due to inclination of the axis of the pumping beam characterized by parameter s (1.71) with $p_0 = 0$, $b = 1$.

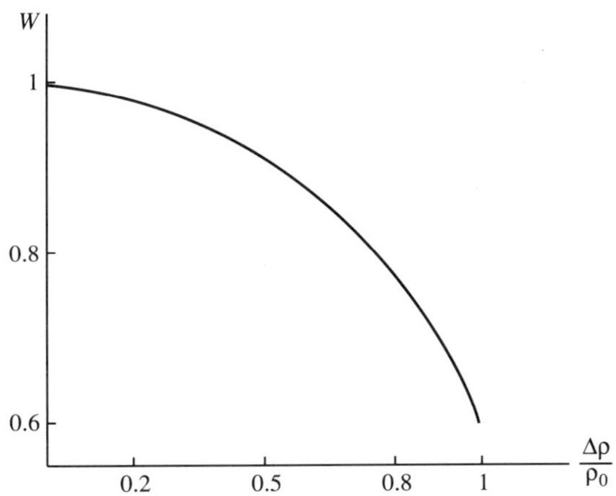

Fig. 1.5 Decrease of dimensionless power W of a trapped TE_{01} mode due to displacement of the axes of the fibre and pumping wave beam $\Delta\rho\rho_0^{-1}$ with $\sigma = 1$, $s = 1$ (eqn (1.71)).

1.2.2 *Spatial modulation of pumping beams by means of computer-generated matched filters*

The modal purity of radiation launched into the fiber may be controlled by suitable spatial modulation of the pumping beam. Such modulation may result, for example, in concentration of light power in one or several rings, consisting of several blobs of light, when the phase takes

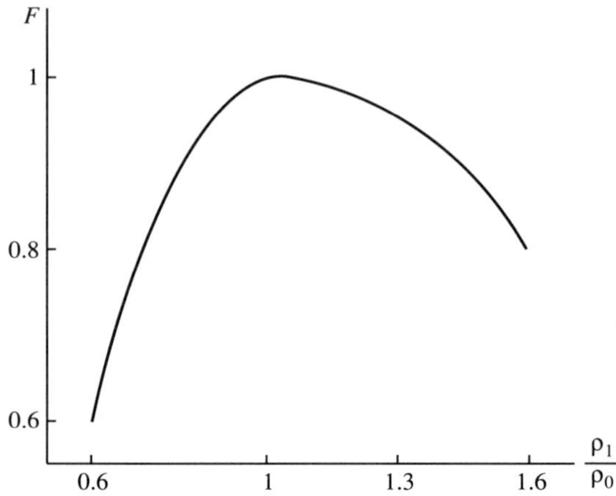

Fig. 1.6 Dependence of power W of the TE_{01} mode excited in a parabolically shaped waveguide upon the ratio of effective radii of pumping wave ρ_1 and trapped mode ρ_0; $u_0 = 0$, $s = 1$.

opposite values from one blob to the next. To illustrate the field configuration, let us look at the input mode filter, consisting of an opaque plate, pierced with a system of holes whose contours follow the level curves of the related mode distribution.

$$|\psi_{\mathrm{pl}}| = \eta|\psi_{\mathrm{pl}}|_{\max}, \qquad \eta < 1$$

(Berdaque and Facq 1982). The perforated plate, covered by glass plates, is set in a collimated beam of monochromatic light. Thus the amplitude of the light field is approximated by a constant value into the holes and zero outside. However, the modal purity of wave beam, passed through this filter, is limited due to formation of parasitic modes as a consequence of diffraction of the beam on these holes. This pernicious trend may be weakened by use of spatial phase filters, similar to computer-generated holograms (Dallas and Lohman 1972). Such filters are acting as sinusoidal phase diffractive gratings with variable depth of modulation.

To illustrate the principles of spatial phase filtration of coherent radiation with wavelength λ_0 in free space, let us consider the phase filter determined by the transmittance function

$$\psi = e^{i\varphi(u,v)}; \qquad \varphi = \pi Q(u,v)\cos\left(\frac{2\pi u}{\delta}\right)\cos\left(\frac{2\pi v}{\delta}\right). \tag{1.73}$$

Here (u, v) are Cartesian coordinates in the filter's plane; δ is the period of spatial modulation of function φ; the function $Q(u, v)$ ($|Q| \leqslant 1$) describes the heterogeneity of the filter's transmittance in the (u, v) plane. This filter is produced by phase relief, manufactured from transparent material with refractive index n. The altitude of this relief,

$$h = \frac{\lambda_0 \varphi(u, v)}{2\pi n}, \qquad (1.74)$$

varies from zero up to maximum value of

$$h_{\text{max}} = \frac{\lambda_0}{2n}. \qquad (1.75)$$

Thus, the maximum phase delay produced by passage of light through this spatial modulator is equal to π. This phase effect may be controlled by means of laser interferometry (Ichioka and Lohman 1972). Such filters are fabricated by reproduction of the profile (1.74) on photographic film, the continuous distribution (1.74) being replaced by numerous small step-like changes of optical density. This quantization, intrinsic to computer-generated profiles of transmittance, is considered to be a source of distortion. To evaluate these distortions, let us represent the field of the incident beam as a superposition of the above-mentioned Gauss–Laguerre eigenmodes (1.67):

$$\psi(u, v) = \sum_{p=1} \xi_p \psi_p(u, v). \qquad (1.76)$$

The passage of this beam through the filter characterized by transmittance function $\psi_p^*(u, v)$, will result in formation of a field $\psi(u, v)$ $\psi_p^*(u, v)$ behind the filter plane, the coefficient ξ_p being determined as

$$\xi_p = \int \psi(u, v) \, \psi_p^*(u, v) \, \mathrm{d}u \, \mathrm{d}v. \qquad (1.77)$$

Focusing of the field $\psi\psi_p^*$ by means of a lens on to the observation plane, coinciding with the waveguide's input, permits excitation of the ψ_p mode in the waveguide (Bartlett *et al.* 1983). The intensity of the excited mode is proportional to $|\xi_p|^2$. However, the real computer-generated filters are distinguished by quantization distortions from ideal ones. Thus, instead of the coefficient ξ_p (1.77), the real filter will produce the coefficient η_p. The difference

$$\varepsilon_p^2 = |\eta_p - \xi_p|^2 \qquad (1.78)$$

may be used for evaluation of modal purity of a synthesized filter. The parameter of importance,

$$\nu_p = \frac{|\varepsilon_p|}{|\xi_p|}, \qquad (1.79)$$

is shown to depend, for given mode, upon the ratio of the beam's radius ρ_0 to the resolution scale σ_0:

$$K_0 = \frac{\rho_0}{\sigma_0}. \tag{1.80}$$

The results of analysis of parameter ν_p (1.79) based on typical parameters for computer-synthesized filters ($\lambda_0 = 0.63\,\mu\text{m}$, $\rho_0 = 6.5 \times 10^{-2}\,\text{cm}$, $\delta = 10^{-2}\,\text{cm}$, $\sigma_0 = 25\,\mu\text{m}$) are illustrated in Fig. 1.7. One can see that the distortions decrease rapidly due to growth of beam's radius, normalized on the resolution scale σ_0; the parameter K_0 (1.80) is equal to $K_0 = 5$; the error ν_p, stipulated by mode quantization, does not exceed 1% of the mode intensity.

Thus, the controlled excitement of modes may be utilized in the design of non-linear waveguide circuits and in evaluation of energy balance in elements of non-linear integrated optics.

1.3 Non-linear phase effects in planar and rectangular waveguides

Utilization of intensity-dependent phase shifts formed in guided wave fields may be considered as one of the key principles of non-linear integrated optics. These shifts are produced by perturbations of the propagation constant β, depending both upon the intensity of trapped radiation and the waveguide geometry and mode of propagation. A way

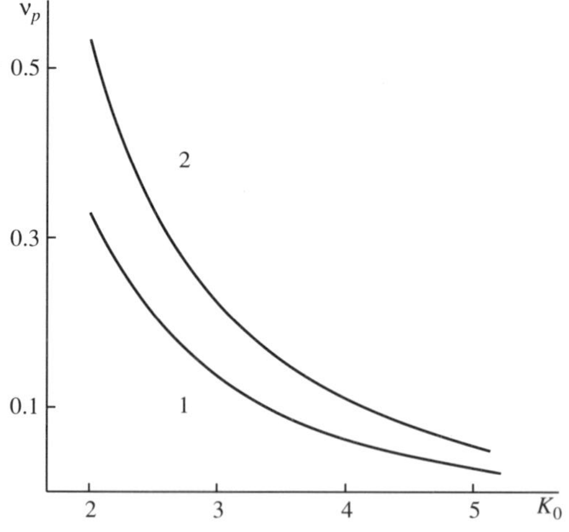

Fig. 1.7 Improvement of modal purity of the TE_{01} (curve 1) and TE_{03} (curve 2) modes formed by computer-synthesized filters and characterized by ν_p values (1.79) due to growth of pumping beam radius ρ_0 normalized on resolution scale $\sigma_0 (K_0 = \rho_0 \sigma_0^{-1})$.

to obtain the generalized form of this dependence is based on use of a formal analogy between the theory of perturbations for Schrödinger's equation in quantum mechanics and the wave equation describing the evolution of an intense wave. In order to apply standard perturbation procedure directly to the present case, let us use the wave equation for guided media:

$$\Delta_1 \psi + [k_\perp^2 - V(x,y)]\psi = 0,$$

$$k_\perp^2 = \frac{\omega^2 n_0^2}{c^2} - \beta^2; \qquad V = \frac{\omega^2}{c^2} F(x,y). \tag{1.81}$$

Variables (x,y) denote here dimensionless coordinates in the waveguide's cross-section; n_0 is the maximum value of the refractive index inside the waveguide; the dimensionless function $F(x,y)$ describes the coordinate dependence of the refractive index. Comparing eqn (1.81) with the stationary Schrödinger equation one can consider the terms ε_0 and V to be similar to 'energy level' and 'potential' respectively. The function $F(x,y)$ may be presented as the sum of a regular part, F_0, determining the waveguide's profile, and a perturbation, F_1, characterizing the non-linearity (Sarid and Stegeman 1981). Perturbation F_1 in a Kerr non-linear medium (1.6) is small, $|F_1| \ll |F_0|$.

According to standard perturbation theory, the perturbation potential

$$V_1 = \frac{\omega^2}{c^2} F_1(x,y) = -\frac{\omega^2}{c^2} \chi \, |\psi(x,y)|^2 \tag{1.82}$$

results in 'energy level' shift

$$\Delta k_\perp^2 = -2\beta \, \Delta\beta. \tag{1.83}$$

The function ψ in eqn (1.81), describing one-directional travelling guided field, may be regarded as a superposition of eigenfunctions of wave equation (1.81). Assuming these eigenfunctions to be orthogonal (1.66), one may obtain from eqns (1.82) and (1.83) the shift of eigenvalue $(\Delta k_\perp^2)_{np}$:

$$(\Delta k_\perp^2) = \frac{\int \psi_{np}^* V_1 \psi_{np} \, dx \, dy}{\int |\psi_{np}|^2 dx \, dy}. \tag{1.84}$$

Integration is performed in (1.84) over the waveguide's cross-section; the variables x and y are normalized on the waveguide's transversal scales. Introducing the change of propagation constant $\Delta\beta_0$ produced by wave field amplitude $\mathbf{E}_0^z$ in bulk homogeneous medium with refractive index n_0,

$$\frac{\Delta\beta_0}{\beta_0} = \frac{\chi|E_0|^2}{2n_0^2}, \tag{1.85}$$

one may rewrite eqn (1.84), using a form that has a simple intuitive meaning:

$$\frac{\Delta\beta_{np}}{\beta_{np}} = \frac{\Delta\beta_0}{\beta_0}\, K_{np}^{-1}\, W_{\perp np}, \tag{1.86}$$

$$K_{np} = 1 - \frac{c^2 k_{\perp np}^2}{\omega^2 n_0^2}, \qquad W_{\perp np} = \frac{\int |\psi_{np}|^4 \, dx\, dy}{\int |\psi_{np}|^2 \, dx\, dy}. \tag{1.87}$$

The dimensionless factors K_{np}^{-1} and $W_{\perp np}$ describe the influence of the waveguide's geometry and mode of propagation, characterized by the quantized spectrum $k_{\perp np}^2$, on the intensity-dependent phase shift φ_{np} accumulated in the course of wave propagation in the npth mode:

$$\varphi_{np} = z\, \Delta\beta_{np}. \tag{1.88}$$

It is important to stress that the aforesaid analysis is valid only for modes characterized by one polarization component of electric field, such as TE_{01} modes in planar and rectangular waveguides and circularly-shaped fibres. Power-dependent phase shifts of travelling fields are determined in each of these cases by a single-valued change of propagation constant $\Delta\beta_{01}$, calculated for the related mode according to eqn (1.86). In contrast, the non-linear evolution of more complicated mode structures containing two or three unequal polarization components may stimulate the appearance of induced gyrotropy. The change of propagation constants $\Delta\beta$ for each polarization component will be different in this case. Such a difference results in intensity-dependent formation of the polarization ellipse and indicates the possibility of governed tuning of polarization structures in waveguide, the non-linearity being caused by the electrostriction effect only ($A = 1$, $B = 0$, Eqn (1.6)).

These phase effects will be discussed below for planar and rectangular waveguides widely utilized in integrated optics. For some practical purposes it seems to be useful to express the change of propagation constant $\Delta\beta$ via the power flow vector $\mathbf{P}$ (Poynting vector),

$$\mathbf{P} = [\mathbf{E} \times \mathbf{H}^*], \tag{1.89}$$

instead of relation (1.85) containing the wave amplitude. In order to avoid the superfluous complication of general formulae produced by this replacement, the related expressions, containing the calculated components of this vector will be analysed further for some simple eigenmodes.

1.3.1 *Power-dependent phase shifts in symmetric planar waveguides*

In order to determine the non-linear changes of propagation constants in planar waveguides it is useful to start from the linear approach to the problem of modes travelling in such structures. Let us discuss the slab shown schematically in Fig. 1.8. The core strip of this structure is assumed to have refractive index n_0; this guiding region is surrounded on both sides by dielectric material with refractive index $n_1 < n_0$ (symmetrical waveguide). Although the waveguide theory of such structures is described in a series of monographs (see, e.g., Marcuse 1974), we shall consider here for future use the trivial TE_{01} mode, which has only one component of electric field E_y and two magentic components H_x and H_z (Fig. 1.8). With $E_x = E_z = H_y = 0$ we obtain from Maxwell's equations (1.1, 1.2) for a travelling wave:

$$\frac{\partial E_y}{\partial x} = i\omega\mu_0 H_z, \tag{1.90}$$

$$-i\beta E_y = i\omega\rho_0 H_x, \tag{1.91}$$

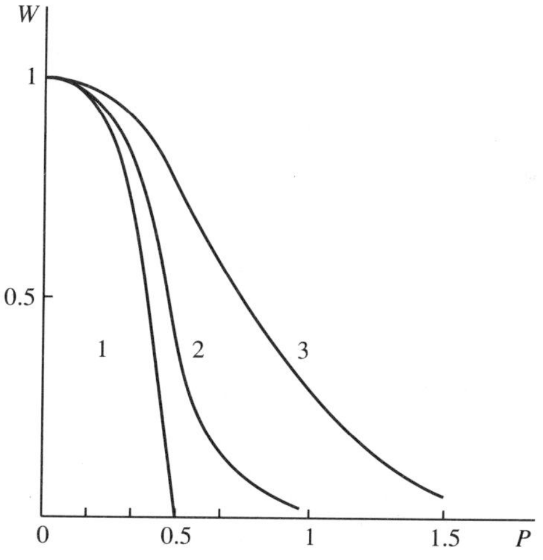

Fig. 1.8 The spatial structure of the TE_{01} mode in the symmetric slab waveguide. The profiles of normalized E_y component $F = E_y E_0^{-1}$, $F|_{x=0} = 1$ are plotted against the normalized distances from the waveguide axis $p = xt^{-1}$; $2t$ is the thickness of central core. The curves 1, 2, and 3 relate to the values of the waveguide parameter γ (1.98) characterizing reflection from the waveguide walls; $\gamma = 1$, 0.8, and 0.5, respectively.

$$i\beta H_x - \frac{\partial H_z}{\partial x} = -i\omega\varepsilon_0 n^2 E_y. \tag{1.92}$$

Substitution of H_x and H_z components expressed in terms of E_y component by means of eqns (1.90) and (1.91) into eqn (1.89) yields the wave equation for the E_y component:

$$\frac{\partial^2 E_y}{\partial x^2} + \left(\frac{\omega^2 n^2}{c^2} - \beta^2\right)E_y = 0. \tag{1.93}$$

The solution of eqn (1.93) must satisfy the boundary conditions at the two dielectric interfaces at $x = \pm t/2$. These conditions require that the tangential E_y and E_z fields be continuous at the dielectric discontinuities situated at the planes $x = \pm t/2$. Solutions that satisfy these conditions for the E_y component and vanish at $x = \pm\infty$ are

$$E_y = E_0 \begin{cases} \cos\left(\dfrac{k_\perp t}{2}\right)e^{-k(x-t/2)}, & x \geqslant t/2; \\[2mm] \cos(k_\perp x), & t/2 \geqslant x \geqslant -t/2; \\[2mm] \cos\left(\dfrac{k_\perp t}{2}\right)e^{k(x+t/2)}; & x \leqslant -t/2. \end{cases} \tag{1.94}$$

Here amplitude factor E_0 depends upon the wave intensity. The E_y component shown by these three expressions satisfies the wave equation (1.93) and is continuous at the dielectric interfaces $x = \pm t/2$. The H_z component is obtained from eqn (1.90). The requirement of continuity of H_z component at the same planes $x = \pm t/2$ leads to the condition

$$k = k_\perp \operatorname{tg}\left(\frac{k_\perp t}{2}\right). \tag{1.95}$$

Combining the formulae for propagation constant β,

$$\beta^2 = \frac{\omega^2 n_0^2}{c^2} - k_\perp^2 = k^2 + \frac{\omega^2 n_1^2}{c^2}, \tag{1.96}$$

with eqn (1.95), we find the eigenvalue equation in the form

$$\frac{\omega^2}{c^2}(n_0^2 - n_1^2) = \frac{k_\perp^2}{\cos^2(k_\perp^t/2)}. \tag{1.97}$$

This leaves $k_\perp$ as the only unknown quantity in eqn (1.97). Here the effective thickness of the guiding layer, depending upon the difference of refractive indices of both media, is about $k_\perp^{-1}$. It is worth emphasizing two limits of this equation. The case $k_\perp t = \pi$ relates to the model of a waveguide with metallic walls, the field being confined totally between the planes $x = \pm t/2$. The opposite case, $k_\perp \to 0$ relates to decrease of the difference $n_1^2 - n_2^2 \to 0$ and to expansion of the guiding layer up to thickness $k_\perp^{-1} \gg t$. To examine all the intermediate cases it is convenient to introduce the dimensionless parameter γ according to the definition

$$k_\perp t = \pi\gamma. \tag{1.98}$$

The values of parameter η for fundamental TE_{01} mode fall in the range $0 \leqslant \gamma \leqslant 1$.

Amplitude E_0 of the trapped radiation may be related to the power carried by the mode. The power is obtained by integrating the z-component of the Poynting vectors (1.89) over the infinite transverse cross section of the waveguide. We thus obtain for the TE_{01} mode

$$P_z = \frac{\beta}{\omega\mu_0} \int_{-\infty}^{\infty} |E_y|^2 \, dx \, dy. \tag{1.99}$$

The evaluation of integral (1.99), the distribution $E_y(x)$ being given by eqn (1.94), results in

$$|E_0|^2 = \frac{2P\omega\mu_0}{\beta l \Delta_0}, \tag{1.100}$$

$$\Delta_0 = (kt)^{-1}\left[kt\left(1 + \frac{\sin k_\perp t}{k_\perp t}\right) + 2\cos^2\left(\frac{k_\perp t}{2}\right)\right]. \tag{1.101}$$

Factor k is defined in (1.95); P/l is energy flow per unit of length in the y-direction; $P = P_z$ (Fig. 1.8), other components of Poynting vector vanishing $(P_x = P_y = 0)$. To simplify the evaluation of $\Delta\beta$ we may redefine the ratio $\Delta\beta\beta^{-1}$ for the TE_{01} mode using the energy flow P/l instead of the square of the amplitude, $|E_0|^2$:

$$\frac{\Delta\beta}{\beta} = \frac{\Delta\beta_0}{\beta_0} K_{01}^{-1} W_{\perp 0}, \tag{1.102}$$

$$\frac{\Delta\beta_0}{\beta_0} = \frac{\chi\omega\mu_0}{2\beta n_0^2} \frac{Pk}{l}. \tag{1.103}$$

Parameter $\Delta\beta_0$ (1.103) relates to non-linear effects in an unbounded homogeneous transparent medium, while the factor $W_{\perp 0}$, calculated from the substitution of field distribution (1.94) into eqn (1.87), illustrates the influence of the mode structure of the guided wave. It is necessary to emphasize that this factor $W_{\perp 0}$ may be considered as a universal function of the parameter γ, defined in (1.98):

$$W_{\perp 0}(\gamma) = \frac{4\left\{\pi\gamma\,\mathrm{tg}\!\left(\dfrac{\pi\gamma}{2}\right)\left[\dfrac{3}{4} + \dfrac{\sin \pi\gamma}{\pi\gamma} + \dfrac{\sin 2\pi\gamma}{8\pi\gamma}\right] + \cos^4\!\left(\dfrac{\pi\gamma}{2}\right)\right\}}{\Delta_1^2},$$

$$\Delta_1 = \pi\gamma\,\mathrm{tg}\!\left(\frac{\pi\gamma}{2}\right)\left[1 + \frac{\sin \pi\gamma}{\pi\gamma}\right] + 2\cos^2\!\left(\frac{\pi\gamma}{2}\right). \tag{1.104}$$

The existence of the limit

$$\lim W_{\perp 0}(\gamma)\big|_{\gamma \to 0} = 1 \tag{1.105}$$

shows that increase in the thickness of waveguide slab will result in continuous transition of factor $W_{\perp 0}$ to its limit value; related to the case of an unbounded medium.

As we pointed out earlier, the value $\gamma = 1$ corresponds to an another important limit of eqn (1.102) connected with presence of metallic planes confined with waveguide slab at the planes $x = \pm t/2$. To analyse the expression (1.102) in this range $(0 < \gamma \leqslant 1)$ it proves convenient to rewrite the right-hand side of eqn (1.102), multiplying both numerator and denominator of this side by the parameter t, which is the slab thickness:

$$\frac{\Delta\beta}{\beta} = GK_{01}^{-1} W_1; \qquad G = \frac{\chi\mu_0\omega}{2\beta n_0^2}\frac{P}{lt}; \tag{1.106}$$

$$W_1 = \pi\gamma\,\mathrm{tg}\!\left(\frac{\pi\gamma}{2}\right) W_{\perp 0}. \tag{1.107}$$

The ratio $P(lt)^{-1}$ in eqn (1.106) determines the density of energy flow through cross section of slab. The graph of the universal function $W_1(\gamma)$ (1.107) is plotted in Fig. 1.9 as well as the graph of another universal function $\theta(\gamma)$, obtained by multiplying the eigenvalue equation (1.97) by the parameter t:

$$\frac{\omega t}{c}\sqrt{(n_0^2 - n_1^2)} = \frac{\pi\gamma}{\cos(\pi\gamma/2)} = \theta(\gamma). \tag{1.108}$$

Plotting both these functions on one figure facilitates the analysis of non-linear phase shifts in the waveguides, the wave frequency ω, the thickness t and refractive indices n_1 and n_2 being arbitrary. Substituting the values of these parameters into the eigenvalue equation (1.108), one may calculate the function $\theta(\gamma)$, and the related value γ_0 is given by the same Fig. 1.9. Parameter K_{01} (1.87), connected with propagation constant β_0, may be rewritten in a form that clarifies the useful dependence of β_0 on slab thickness t:

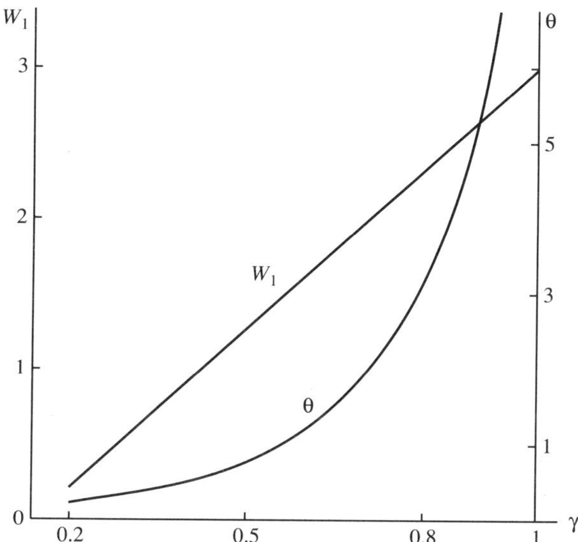

Fig. 1.9 Graphs of the universal functions $W_1(\gamma)$ (1.107) and $\theta(\gamma)$ (1.108) illustrating the influence of partially reflecting waveguide walls described by parameter γ (1.98) on the non-linear perturbation of propagation constant β.

$$K_{01} = 1 - \left(\frac{c\pi\gamma}{\omega n_0 t}\right)^2. \tag{1.109}$$

Combining eqn (1.86) and (1.88) one may finally determine the non-linear phase shift φ_{01}.

Such intensity-dependent phase phenomena are important for the series of power-carrying circuits in integrated optics. Thus, an actual example is connected with stripe heterostructures, used in double-heterostructure injection lasers (Chandra *et al.* 1981). A typical structure of this kind contains central GaInAsP film surrounded on both sides by InP layers (Fig. 1.10). These systems are usually operated in the spectral range about $1\,\mu\text{m} < \lambda < 1.7\,\mu\text{m}$ (Reinhart *et al.* 1971). The non-linear factor W_1 (1.107), calculated for the wavelengths in this spectral range, is presented in Fig. 1.11 for different thicknesses of guiding slabs supporting TE_{01} modes.

It is worthwhile concluding this section with an important remark. Examination of function W_1 (1.107) shows the existence of the limit

$$\lim W_1|_{\gamma=1} = 3 \tag{1.110}$$

related to the model of totally reflecting planes confined the transparent medium, $-t/2 \leqslant x \leqslant t/2$. The universal function W_1 (1.107) is

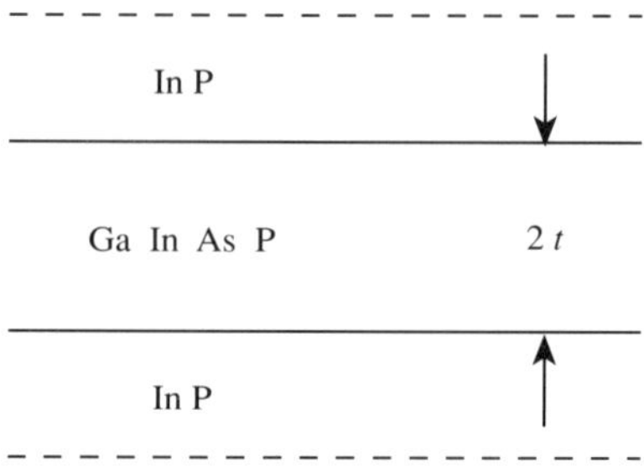

Fig. 1.10 Double heterostructure containing a central GaInAsP film surrounded by InP layers.

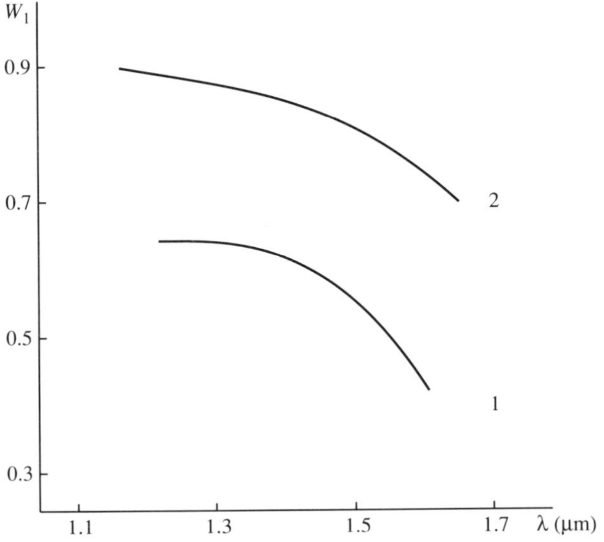

Fig. 1.11 The non-linear factor W_1 (1.107) plotted against the wavelengths λ (μm) of guided radiation in heterostructures with thickness of central film $t = 0.4\,\mu$m (curve 1) and $t = 0.6\,\mu$m (curve 2).

continuous in the range $0 \leqslant \gamma \leqslant 1$; thus one may evaluate the tendency for non-linear evolution of the guided mode in the dielectric waveguide, related to the case $\gamma \leqslant 1$, using the limit $\gamma = 1$. The calculations are simplified considerably, and the errors in the range, for example, $0.8 < \gamma < 1$, do not exceed 20%. Therefore, using the limit $\gamma \to 1$ proves to be a suitable approximation for understanding of series of non-linear effects in guided systems. This approximation will be used further in order to avoid the rigorous computational problems in discussion of physically meaningful phenomena.

1.3.2 *Geometrical effects in self-induced gyrotropy of rectangular waveguides*

Non-linear propagation of guided fields containing two or three components of polarization such as the higher TE and TM modes is characterized by a peculiar phase effect, which is distinguished in principle from the above-mentioned phase shift of the TE_{01} mode. Discussing, for definitness, the TE_{11} mode in a rectangular waveguide in the low-intensity limit, we consider both E_x and E_y components of the electric field of the trapped wave as travelling with the same phase velocity, determining by propagation constant β. In contrast, the analysis of intensity-dependent perturbations of propagation constants for this waveguide shows these perturbations to be different for E_x and E_y components. This inequality indicates non-linear splitting of the travelling mode for two modes propagating with slightly different phase velocities. Such splitting results in self-induced gyrotropy of the waveguide medium and to formation of tunable polarization ellipses in the waveguide cross section. This effect will be shown to depend upon the ratio of the scales of waveguide rectangular cross section. This geometrical effect vanishes when the scales are equal. To examine this possibility of the tuning of modes' polarization structure, we shall use the model of a waveguide with totally reflecting walls, related to the limit $\gamma = 1$ (1.98).

In order to evaluate this non-linear polarization tuning of the TE_{11}, mode we must reconsider the traditional analysis of the polarization structure of this mode. Keeping in mind the rectangular waveguide shown in Fig. 1.12, let us use the wave equation for TE modes, derived straightforwardly from Maxwell's equations (1.1) and (1.2) and governing the amplitude of longitudinal component of wave's magnetic field, H_z:

$$\frac{\partial^2 H_z}{\partial x^2} + \frac{\partial^2 H_z}{\partial y^2} + k_\perp^2 H_z = 0. \tag{1.111}$$

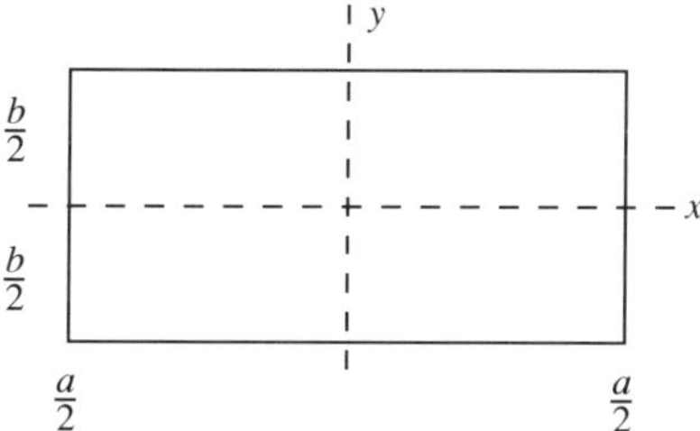

Fig. 1.12 Cross-section of the rectangular waveguide with $u_0 = 4$.

Here $k_\perp$ is the transverse component of the wave vector determined according to (1.61). The components of the travelling wave field in the TE mode are known to be expressed in terms of H as

$$E_x = -\frac{i\mu_0\omega}{k_\perp^2}\frac{\partial H_z}{\partial y}; \qquad E_y = +\frac{i\mu_0\omega}{k_\perp^2}\frac{\partial H_z}{\partial x}; \qquad E_z = 0; \quad (1.112)$$

$$H_x = -\frac{\beta}{\mu_0\omega}E_y; \qquad H_y = \frac{\beta}{\mu_0\omega}E_x. \qquad (1.113)$$

The components E_x and E_y obey to the boundary conditions

$$E_x = \left|\begin{matrix} y = \dfrac{b}{2} \\[2mm] \\[2mm] y = -\dfrac{b}{2} \end{matrix}\right. = 0; \qquad E_y = \left|\begin{matrix} x = \dfrac{a}{2} \\[2mm] \\[2mm] x = -\dfrac{a}{2} \end{matrix}\right. = 0 \qquad (1.114)$$

Choosing the solution of eqn (1.111) in the form

$$H_z = \mathrm{C}\sin k_\perp x \sin k_\perp y \qquad (1.115)$$

we may obtain the formulae for E_{x0} and E_{y0} in the low-intensity limit:

$$E_{x0} = \frac{Ak_2}{k_\perp}\sin k_\perp x \cos k_2 y; \qquad E_{y0} = -\frac{Ak_1}{k_\perp}\cos k_1 x \sin k_2 y,$$

$$A = -i\omega\mu_0 k_\perp^{-1}C. \qquad (1.116)$$

Using boundary conditions (1.114) yields the following values of parameters k_1 and k_2:

$$k_1 = \frac{\pi}{a}; \qquad k_2 = \frac{\pi}{b}; \qquad k_1^2 + k_2^2 = k_\perp^2. \qquad (1.117)$$

Comparison of eqn (1.117) with eqn (1.98) shows, that we are dealing with the case $\gamma = 1$, related to the model of totally reflecting waveguide walls. The unknown constant A (1.116) is connected with energy flow P (1.89). Substitution of relations (1.113) into eqn (1.89) and integration over the waveguide cross section leads to the result

$$A = \left(\frac{2P}{ab}\frac{\mu_0\omega}{\beta}\right)^{1/2}. \qquad (1.118)$$

Now we may start the analysis of the non-linear evolution of the TE_{11} mode. At the beginning it is worth reconsidering the approach developed in the previous section. In fact, such approach is valid as long as the perturbation of propagation constant $\Delta\beta$ is supposed to be invariable in the course of propagation: however, we shall face a series of non-linear problems where this simple supposition becomes invalid. Therefore, it is useful to generalize this approach in order to eliminate the aforesaid restriction. An example of such a generalized approach in application to the TE_{11} mode is developed further.

Let us represent the components E_x and E_y by means of new functions $F_1(z)$ and $F_2(z)$:

$$E_x(z) = E_{x0}F_1(z); \qquad E_y(z) = E_{y0}F_2(z). \tag{1.119}$$

Here $F_1(z)$ and $F_2(z)$ are slowly varying complex dimensionless functions; the functions E_{x0} and E_{y0} are determined in (1.116). Substitution of eqn (1.119) into the general non-linear equation (1.17) results in the appearance of a pair of coupled equations describing the evolution of functions $E_x(z)$ and $E_y(z)$. Multiplication of these equations with the eigenfunctions related to E_x and E_y, respectively, and integration of these products over the waveguide cross-section provides equations governing the functions F_1 and F_2. It is convenient to use the dimensionless variable τ:

$$\tau = \frac{z}{L_0}; \qquad L_0 = \frac{16\lambda_0 n_0 \sqrt{1 + y^2}}{\pi\chi|A|^2}. \tag{1.120}$$

λ_0 is the free-space wavelength and n_0 is the refractive index of waveguide medium. Now one may find the dimensionless equations of the non-linear evolution of the TE_{11} mode, in the form

$$i\frac{\partial F_1}{\partial \tau} + F_1(9y_2^2|f_1|^2 + y_1^2|f_2|^2) = 0, \tag{1.121a}$$

$$i\frac{\partial F_2}{\partial \tau} + F_2(y_2^2|f_1|^2 + 9y_1^2|f_2|^2) = 0. \tag{1.121b}$$

Here $y_{1,2}$ denote the dimensionless parameters

$$y_{1,2} = \frac{k_{1,2}}{\beta}. \tag{1.122}$$

Let us by analogy with eqn (1.27) represent the functions F_1 and F_2 in eqns (1.121) as the products of 'amplitude' and 'phase' factors:

$$F_1 = f_1 e^{i\varphi_1}; \qquad F_2 = f_2 e^{i\varphi_2}; \qquad f_{1,2} = |F_{1,2}|. \tag{1.123}$$

Substitution of (1.123) into system (1.121) shows that the amplitudes f_1 and f_2 remain constant in the course of such self-action of the TE_{11} mode, while the phase shift between E_y and E_x components,

$$\varphi = \varphi_2 - \varphi_1, \tag{1.124}$$

varies uniformly from the initial value $\varphi|_{\tau=0} = 0$,

$$\varphi_{TE} = 8y_1^2(1 - u_0)\tau. \tag{1.125}$$

Here the parameter of importance, u_0, is connected with the waveguide geometry:

$$u_0 = \frac{y_2^2}{y_1^2} = \left(\frac{a}{b}\right)^2. \tag{1.126}$$

Thus, the phase shift φ is shown to be positive or negative depending on the ratio of the scales of waveguide cross section, $a < b$ or $a > b$, respectively (Fig. 1.13). This phase shift caused by the difference of scales is vanishing the cross section being square (a = b).

The appearance of non-linear phase shift φ between the components of polarization stipulates the formation of new polarization structures of the TE_{11} mode. Unlike the low-intensity limit, related to linear polarization of the wave field (1.116), the uniformly varying phase φ (1.125) provides the formation of elliptically polarized waves; the shape and orientation of this ellipse changes continuously in the course of mode propagation. Using the coordinate system shown in (Fig. 1.12) we may derive the expressions for the major E_1 and minor E_2 axes of this ellipse in the form (Fig. 1.14)

$$E_{1,2} = \frac{Ak_2}{k_\perp} F_{1,2}; \qquad F_{1,2}^2 = \frac{1}{2}(1 + u_0^{-1} \pm R); \tag{1.127}$$

$$R = \sqrt{(1 + 2u_0^{-1}\cos 2\varphi + u_0^{-2})}. \tag{1.128}$$

The angle θ between the major axis F_1 and the x-direction depends pon the parameter R (1.128):

$$\theta = \frac{1}{2} \arccos\left(\frac{1 - u_0^{-1}}{R}\right). \tag{1.129}$$

The monotonic non-linear variation of phase shift (1.125) leads to oscillations of angle θ (1.129) between the values $\theta = \pm\theta_1$ (Fig. 1.15):

$$\theta_1 = \frac{1}{2} \arccos\left(\frac{1 - u_0^{-1}}{R}\right). \tag{1.130}$$

The configurations $u_0 > 1$ and $u_0 < 1$ relate to turning of the ellipse in the clockwise and counter-clockwise direction, respectively. The spatial period of such oscillations may be derived from eqn (1.125):

$$\tau = \frac{\pi}{4y_1^2|u_0 - 1|}. \tag{1.131}$$

The linear polarization ($\varphi = \pi p$) is formed at the points $\tau = \pi p[8y_1^2|u_0 - 1|]^{-1}$, p being integer. The elliptical polarization exists

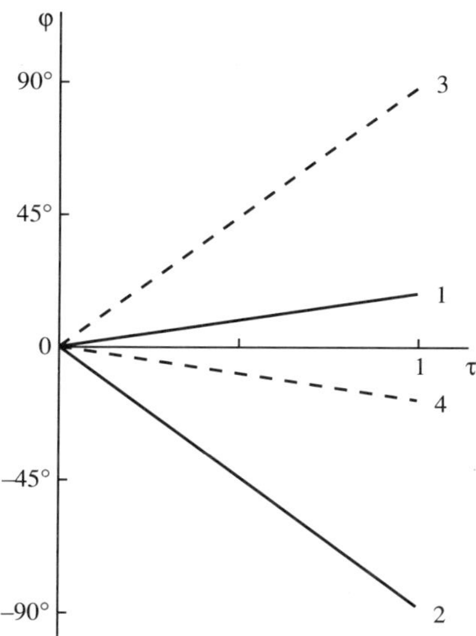

Fig. 1.13 Geometric effect in change of sign of phase shift σ between the components of the TE_{11} mode in a rectangular waveguide. The curves 1 and 2 relate to the ratio of waveguide sides $ab^{-1} = \sqrt{u_0} = 2$ and 0.5, respectively, τ is the normalized propagation distance.

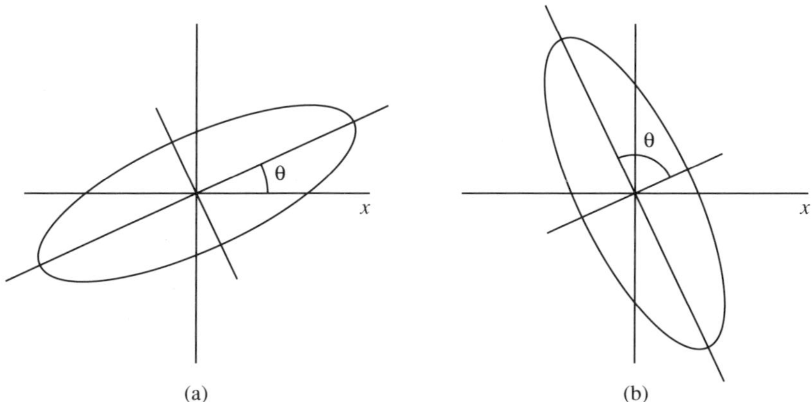

Fig. 1.14 The influence of waveguide geometry on self-induced gyrotropy of the TE_{11} mode: (a) and (b) illustrate the different orientations of the polarization ellipses in the cases $u_0 = 4$ and $u_0 = 0.25$, respectively, at positions τ related to half-periods of the non-linear oscillations of the polarization in each case.

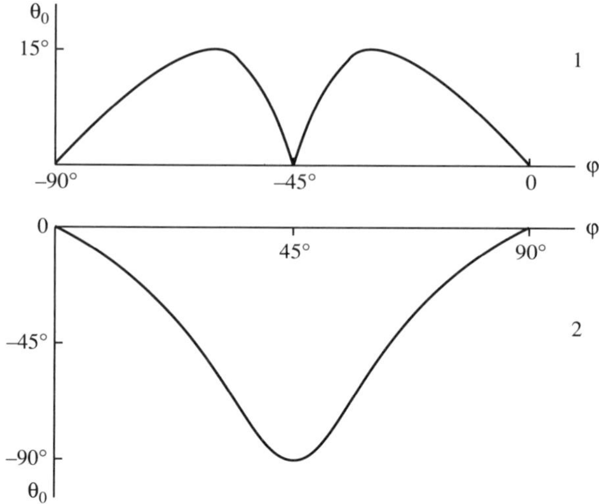

Fig. 1.15 Non-linear oscillations of the angle θ between the ellipse's major axes and the direction of linear polarization of the TE_{11} mode at the waveguide input $\tau = 0$. Curves 1 and 2 correspond to the cases $u_0 = 4$ and $u_0 = 0.25$, respectively.

at intermediate points: the maximum eccentricity achieved at the points $\tau = \pi(2p + 1)\,[\,16y_1^2|u_0 - 1|\,]^{-1}$ is

$$\frac{F_1}{F_2} = (1 + u_0)^{-1/2}. \tag{1.132}$$

Before discussion of peculiarities of this intensity-dependent polarization of the TE_{11} mode, let us illustrate briefly the similar effect for the TM_{11} mode too. Starting with the E_x and E_y components of this mode described in the framework of low-intensity limit by expressions (Marcuse 1974) satisfying boundary conditions (1.114),

$$E_{x0} = A\,\frac{k_1}{k_\perp}\,\sin k_1 x \cos k_2 y; \qquad E_{y0} = A\,\frac{k_2}{k_\perp}\,\cos k_1 x \sin k_2 y, \tag{1.133}$$

we may follow the scheme of self-action analysis shown above for the TE_{11} mode. Using the same normalized distance of self-action τ and the notations (1.122–1.124) we conclude that the amplitudes of polarization components $f_{1,2}$ (1.123) remain invariable again; the self-action leads to formation of non-linear phase shift φ between the E_y and E_x components of the TM_{11} mode (Fig. 1.13):

$$\varphi_{TM} = 8y_1^2(u_0 - 1)\tau. \tag{1.134}$$

Comparison of phase shifts for the TM_{11} and TE_{11} (1.125) modes shows that these shifts have opposite signs,

$$\varphi_{TM} = -\varphi_{TE}, \qquad (1.135)$$

the modes' power being the same. Analogously to self-action of the TE_{11} mode, the polarization ellipse forms in the course of propagation of the TM_{11} mode too: however, the directions of turning of these ellipses are opposite for TE_{11} and TM_{11} modes travelling in the same waveguide.

The appearance of phase shifts between the components of polarization of a powerful wave may be treated as a manifestation of self-induced birefrigence and gyrotropy. We emphasize three important results in regard to these effects in rectangular waveguides, the waveguide material itself being isotropic.

1. In contrast to the non-linear rotation of the polarization ellipse in an unbounded medium, produced by power-dependent gyrotrophy of these media and described by the term $B(\mathbf{E})^2\mathbf{E}^*$ in non-linear polarization (1.6), we are dealing here with the electrostriction mechanism of non-linearity ($B > 0$). This case is important, for example, for silica fibers, widely used in integrated optics (Menyuk 1989).

2. Instead of invariance of the shape of the initial polarization ellipse in the course of its uniform non-linear rotation, described by eqn (1.50), the shape of ellipse formed in the waveguide under discussion varies, starting with the linear polarization at the beginning of evolution ($\varphi = 0$). Here both the shape of this ellipse and the inclination of its major, axis oscillate simultaneously between two limit states related to linear polarizations.

3. The spatial periods of oscillation of mode parameters, deriving from intensity-dependent birefringence, depend upon the ratio of the waveguide's cross section sides (a/b). The role of this geometric effect is clearly manifested in the vanishing of the above-mentioned oscillations in a square-shaped waveguide ($a = b, u_0 = 1$).

The level of power necessary for the development of such non-linear polarization may be evaluated by means of eqns (1.118) and (1.120). To observe these effects, e.g. the formation of the polarization ellipse with eccentricity (1.132) from initial linear polarization, in the span of a waveguide with $a = 0.5\,\mu\text{m}$, $b = 1\,\mu\text{m}$, $u_0 = 0.25$, $l = 1\,\text{m}$, $\lambda_0 = 1.06\,\mu\text{m}$, the transmitted power must be about 1–2 W.

The non-linear interaction of components of polarization described above results in phase-modulation effects, the amplitudes of these components remaining constant. This conclusion is valid as long as the frequency of the travelling mode is considered to be far from the cutoff frequency of the waveguide. However, the interaction of polarization components in the vicinity of the cutoff frequency will be shown to provide the coupled amplitude–phase self-modulation of trapped modes.

1.4 Amplitude–phase self-modulation of travelling modes close to cutoff

The mode structure of the trapped field in the waveguide results in formation of spatial scales characterizing the heterogeneous distribution of this field in the waveguide's cross section. Simple examples of such transversal structure were given, for example, by radial envelopes of TE_{0p} modes (Fig. 1.3). These transversal scales of mode structure are smaller than the characteristic scales of the heterogeneities of the refractive index forming the guiding channel; thus one may obtain for TE_{0p} modes that $\rho_0 < a$ (1.64). The existence of these scales liduces the appearance of peculiar spatial dispersion of the non-linear susceptibility of Kerr-law dielectrics. This effect is described by eqn (1.3); rewritten in the form

$$\Delta \mathbf{E} = - \frac{\chi(\mathbf{E}\nabla|\mathbf{E}|^2)}{\mathscr{E}} \tag{1.136}$$

$\mathscr{E}$ is the dielectric susceptibility in the low-intensity limit. The previous analysis of non-linear evolution of guided modes was based on div $\mathbf{E} = 0$ following from eqn (1.8) with $\mathscr{E} = $ constant. The deviation of div $\mathbf{E}$ from zero value is produced by Kerr non-linearity of the guiding medium (1.136); this effect must be taken into account alongside the dependence of polarization upon the wave's intensity (1.6).

Rigorously speaking, eqn (1.17), governing the self-action of trapped modes, must be reconsidered in order to include the hitherto ignored term div $\mathbf{E}$. Reinstating this omitted term in the general equation (1.18) by means of eqn (1.136), we obtain instead of eqn (1.17) the generalized equation for the evolution of a continuous powerful wave in a waveguide:

$$2i\beta \frac{\partial E_i}{\partial z} + \chi\left[\left(\frac{\mathbf{E}_j \nabla_j|\mathbf{E}|^2}{\mathscr{E}}\right) + \frac{\omega^2}{c^2}\left(A|\mathbf{E}|^2 E_j + BE^2 E_j^*\right)\right] = 0. \tag{1.137}$$

The influence of the wave's intensity on its self-action is described in eqn (1.137) by two terms, related to local and non-local non-linear effects, respectively; the non-local effect connected with derivatives $\nabla\mathbf{E}$. Therefore, the contribution of non-local effects is evaluated by factors

$k_\perp \beta^{-1}$ or $k_\perp^2 \beta^{-2}$ in comparison with the contribution of local effects to the total non-linear susceptibility. The above-mentioned analysis was limited to the case $k_\perp \beta^{-1} \ll 1$, which relates to a waveguide regime far from cutoff. The increase of the contribution of non-local effects ($k_\perp \leqslant \beta$) relates to the approach of the waveguide regime to cutoff. The approach of the waveguide regime to cutoff, the condition $k_\perp \beta^{-1} \leqslant 1$ being fulfilled, indicates the increase of the contribution of non-local effects. Thus some new non-linear phenomena involving by non-vocal effects are expected to exist close to cutoff. The basic features of the theory of self-modulation of guided modes, with arbitrary ratio $k_\perp \beta^{-1}$, are considered here.

First of all let us mention some cases where the manifestation of effects of $k_\perp \beta^{-1}$ order is limited owing to the geometry of the problem.

1. TE_{0p} modes in planar, rectangular and circularly-shaped waveguides; the non-local effect vanishes rigorously owing to the condition $(E_j \nabla_j |\mathbf{E}|^2) = 0$.

2. The TM_{01} mode is characterized by increase in non-linear phase shift due to the non-local effect, while the amplitude modulation proves to be negligible. It is worth discussing this case in detail in order to present a scheme of analysis for such phenomena.

Let us consider the propagation of the TM_{01} mode in a planar slab using for simplicity, by analogy with Section 1.3.2, the model of totally reflecting planes located at the points $x = \pm a/2$. This mode contains E_x and E_z components of the electric field ($E_y = 0$, Fig. 1.16):

$$E_{x0} = M \sin k_\perp x; \qquad E_{z0} = ik_\perp \beta^{-1} M \cos k_\perp x. \qquad (1.138)$$

Introducing by analogy with eqn (1.119) the dimensionless, slowly varying real functions $f_1(z)$, $f_2(z)$, $\varphi_1(z)$ and $\varphi_2(z)$,

$$E_x = E_{x0} f_1 e^{i\varphi_1}; \qquad E_z = E_{z0} f_2 e^{i\varphi_2}, \qquad (1.139)$$

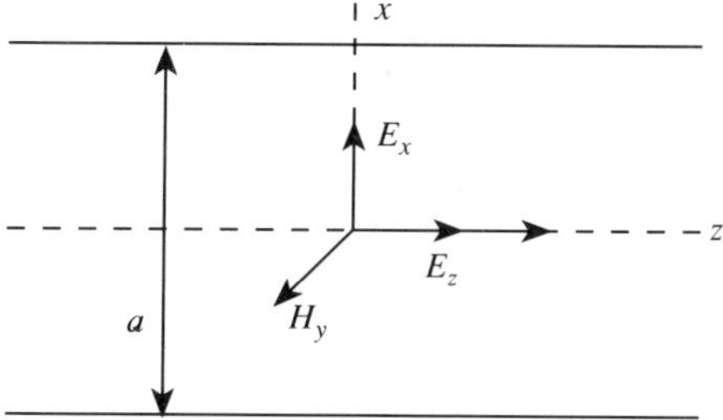

Fig. 1.16 Polarization structure of the TM_{01} mode propagating in the z-direction in a slab waveguide.

we may present the intensity-dependent part of the susceptibility of the guiding slab $\mathscr{E}_2$, supposing for definiteness that $B = 0$ (1.137), in the form

$$\mathscr{E}_2(x) = \mathscr{E}_0 \chi |M|^2 [f_1^2 \sin^2(k_\perp x) + y^2 f_2^2 \cos^2(k_\perp x)]; \quad (1.140)$$

$$y^2 = k_\perp^2 \beta^{-2}; \qquad k_\perp = \pi a^{-1}. \qquad (1.141)$$

Parameter y (1.141) is shown to be the factor of importance for further analysis. Making use of distribution (1.140) and eqn (1.136) we may calculate the divergence of the electric field:

$$\mathbf{E} = -2\chi \mathscr{E}^{-1} |M|^2 M k_\perp f_1 \sin^2(k_\perp x) \cos(k_\perp x)(f_1^2 - y^2 f_2^2) e^{i\beta z}. \qquad (1.142)$$

Substitution of expressions (1.140) and (1.142) into eqn (1.137) leads to a set of equations depending upon the variables x and z. Elimination of the dependence upon the transversal x-coordinate by the integration of the equations obtained, using the eigenfunctions $\sin(k_\perp x)$ and $\cos(k_\perp x)$, finally yields the following system of equations for determination of the dimensionless amplitudes f_1, f_2 and phases φ_1 and φ_2:

$$\frac{\partial f_1}{\partial \tau} = 0; \qquad \frac{\partial f_2}{\partial \tau} = 2(f_1^2 - y^2 f_2^2) \sin \varphi;$$

$$\frac{\partial \varphi_1}{\partial \tau} = -2y^2 (f_1^2 - y^2 f_2^2) + (1 + y^2)(3f_1^2 + y^2 f_2^2); \qquad (1.143)$$

$$f_2 \frac{\partial \varphi_2}{\partial \tau} = 2f_1(f_1^2 - y^2 f_2^2) \cos \varphi + f_2(1 + y^2)(f_1^2 + 3f_2^2).$$

The normalized distance τ in (1.143) is defined as

$$\tau = \frac{z}{L_0}; \qquad L_0 = \frac{4\lambda_0 n_0 \sqrt{1 + y^2}}{\pi \chi |M|^2}. \qquad (1.144)$$

Here and further we use the identity

$$\frac{\omega^2 n^2}{c^2 \beta^2} = 1 + y^2.$$

The solution of system (1.143) must obey the initial conditions

$$f_{1,2}\big|_{\tau = 0} = 1; \qquad \varphi_{1,2}\big|_{\tau = 0} = 0. \qquad (1.145)$$

Introducing the phase shift between E_z and E_x components,

$$\varphi = \varphi_2 - \varphi_1,$$

it is easy to find two invariants of the system (1.143):

$$f_1 = 1; \qquad f_2^2 - 2f_2 \cos \varphi = -1.$$

The last invariant exists only for the values $f_2 = 1$, $\varphi = 0$. Thus the dimensionless amplitudes f_1 and f_2, as well as the phase shift between them, are invariant in the course of non-linear evolution. Here the power-dependent phase shifts $\varphi_{1,2}$ of each component E_x and E_z vary according to equation, appearing from system (1.143),

$$\frac{\partial \varphi_{1,2}}{\partial \tau} = F(y^2) = 3 + 2y^2 + 3y^4. \qquad (1.146)$$

The graphs depicted in Fig. 1.17 illustrate the considerable growth of phase shifts $\varphi_{1,2}$ even in the case of moderate values of parameter $y \leqslant 1.25$. The relative phase shift is absent: $\varphi_1 = \varphi_2$.

Although the dimensionless function f_2, connected with normalized longitudinal component E_z, remains constant, this component E_z is itself shown by eqn (1.142) to be renormalized slightly by means of an additional non-linear correction:

$$E_z = E_{zo}[1 - 2\chi|M|^2 \mathscr{L}^{-1}(1 - y^2)]. \qquad (1.147)$$

One may see that the order of magnitude of this correction is determined by small parameter $\chi|M|^2 \ll 1$.

We emphasize two important results of this analysis;

1. Polarization structures of both fundamental TM_{01} and TE_{01} modes remain invariant in the course of non-linear propagation, and self-induced birefringency does not appear.

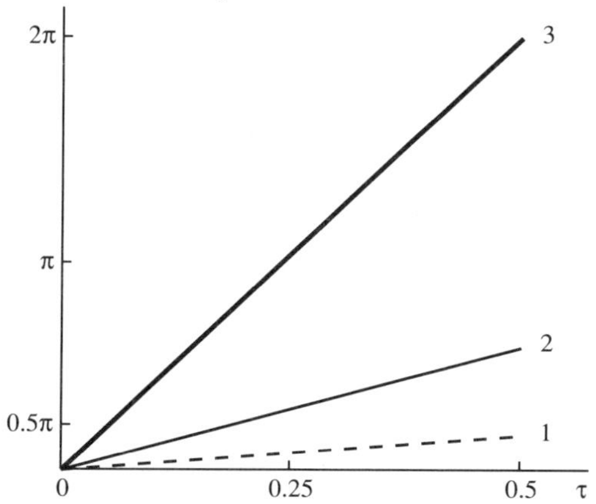

Fig. 1.17 The increase of non-linear phase φ of the TM_{01} mode in a slab waveguide near the cutoff versus the distance τ. Curves 1, 2, and 3 relate to the values $y^2 = k_1^2 \beta^{-2} = 0$, 1 and 1.5, respectively.

2. Intensity-dependent phase shifts of both the E_x and E_z components of TM_{01} mode, unlike the case for TE_{0p} modes, depend essentially upon the contribution of non-local non-linear effects (Fig. 1.17).

The present discussion is limited to the case of fundamental TE_{01} and TM_{01} modes. Proceeding in a similar fashion, we may analyse the self-action of the higher modes, distinguished by new trends in non-linear evolution. These trends are analysed below by means of simple TE_{11} and TM_{11} modes.

1.4.1 *Non-linear splitting of the TE_{11} mode*

Starting with the polarization structure of the TE_{11} mode in a rectangular waveguide given above by eqn (1.116), we may obtain the power-dependent part of dielectric susceptibility in the form

$$\mathscr{E}_2 = \mathscr{E}_0 \chi |A|^2 \beta^2 [y_2^2 \sin^2 k_1 x \cos^2 k_2 y + y_1^2 \cos^2 k_1 x \sin^2 k_2 y].$$

$$(1.148)$$

This expression was used in Section 1.3.2. for analysis of phase self-modulation provided by local non-linear effects. To find the non-local effect, expression (1.148) must be substituted into eqn (1.137). Gathering the results and omitting the intermediate calculations, we obtain the system of equations governing the self-action of components $F_{1,2} = f_{1,2} \exp(i\varphi_{1,2})$ of the TE_{11} mode:

$$i\frac{\partial F_1}{\partial \tau} + F_1[(1 + y^2)(9y_2^2 f_1^2 + y_1^2 f_2^2) + 2y_1^2(y_1^2 f_2^2 - 3y_2^2 f_1^2)]$$

$$+ 2y_1^2 F_2(3y_1^2 f_2^2 - y_2^2 f_1^2) = 0; \qquad (1.149)$$

$$i\frac{\partial F_2}{\partial \tau} + F_2[(1 + y^2)(y_2^2 f_1^2 + 9y_1^2 f_2^2) + 2y_2^2(3y_1^2 f_2^2 - y_2^2 f_1^2)]$$

$$+ 2y_2^2 F_1(y_1^2 f_2^2 - 3y_2^2 f_1^2) = 0. \qquad (1.150)$$

The values of parameters $y_{1,2}^2$ determining the approach of the regime of TE_{11} mode propagation to cutoff are arbitrary in the system (1.149, 1.150). The variable τ is defined as

$$\tau = zL_{TE}^{-1}; \qquad L_{TE} = \frac{16\lambda_0 n_0 \sqrt{1 + y^2}}{\pi \chi |A|^2}. \qquad (1.151)$$

The limit $y_{1,2}^2 \ll 1$ relates to the reduced equations (1.120, 1.121) analysed above.

Basing on these generalized equations, we may advance the theory of coupled amplitude–phase self-modulation of waves travelling in the TE_{11} mode. Unlike the reduced system (1.121) limited to the case of self-action of each component of polarization, the system obtained describes simultaneously the effects of both self-action and interaction of components E_x and E_y. The main tendencies of these self-modulation processes may be analysed by means of invariants of the system (1.149, 1.150). While keeping the initial conditions (1.123) and introducing the functions

$$u = u_0 f_1^2 f_2^{-2}, \qquad \varphi = \varphi_2 - \varphi_1, \tag{1.152}$$

with u_0 given in (1.126), we may write this invariant in the form

$$\cos\varphi = \frac{1}{2y^2\sqrt{uu_0}} \left\{ y_1^2(1 + uu_0) - (1 - u)\left[1 + y^2 - (1 + 2y_2^2) \right.\right.$$
$$\left.\left. \left(\frac{uu_0 - 1 + u_0 - u}{uu_0 - 1 - u_0 + u}\right)^{2/3} \right] \right\}. \tag{1.153}$$

The initial value $u = u_0$ relates to the case $\varphi = 0$.

The behaviour of the phase shift φ between E_y and E_x, for a given component waveguide geometric parameter $u_0 = a^2 b^{-2}$, depends upon the parameter $y_1^2 = k_1^2\beta^{-2}$ connected with the wave's frequency ω; here parameters y_2^2 and y^2 are expressed as

$$y_2^2 = u_0 y_1^2; \qquad y^2 = y_1^2(1 + u_0) \tag{1.154}$$

Invariant (1.153) indicates the period oscillations of parameter u (1.151) between the limits $u = u_0$ and $u = u_1(y_1^2)$; the related values of phase shift φ vary between $\varphi = 0$ and $\varphi = \varphi_1(y_1^2)$. The tendencies of these variations are illustrated in Fig. 1.18. The phase shift φ varyies monotonically from $\varphi = 0$ up to $\varphi_1 = \pm\pi$; both of these limits relate to linear polarization.

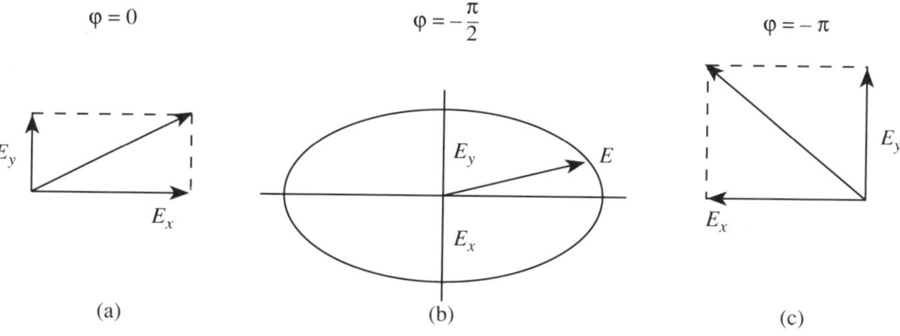

Fig. 1.18 Non-linear oscillations of the polarization structure of the TE_{11} mode in a rectangular waveguide ($u_0 = 4$) between the values $\varphi = 0$ (a) and $|\varphi| = \pi$ (c), related to linear polarization. The polarization at intermediate points (b) is elliptical.

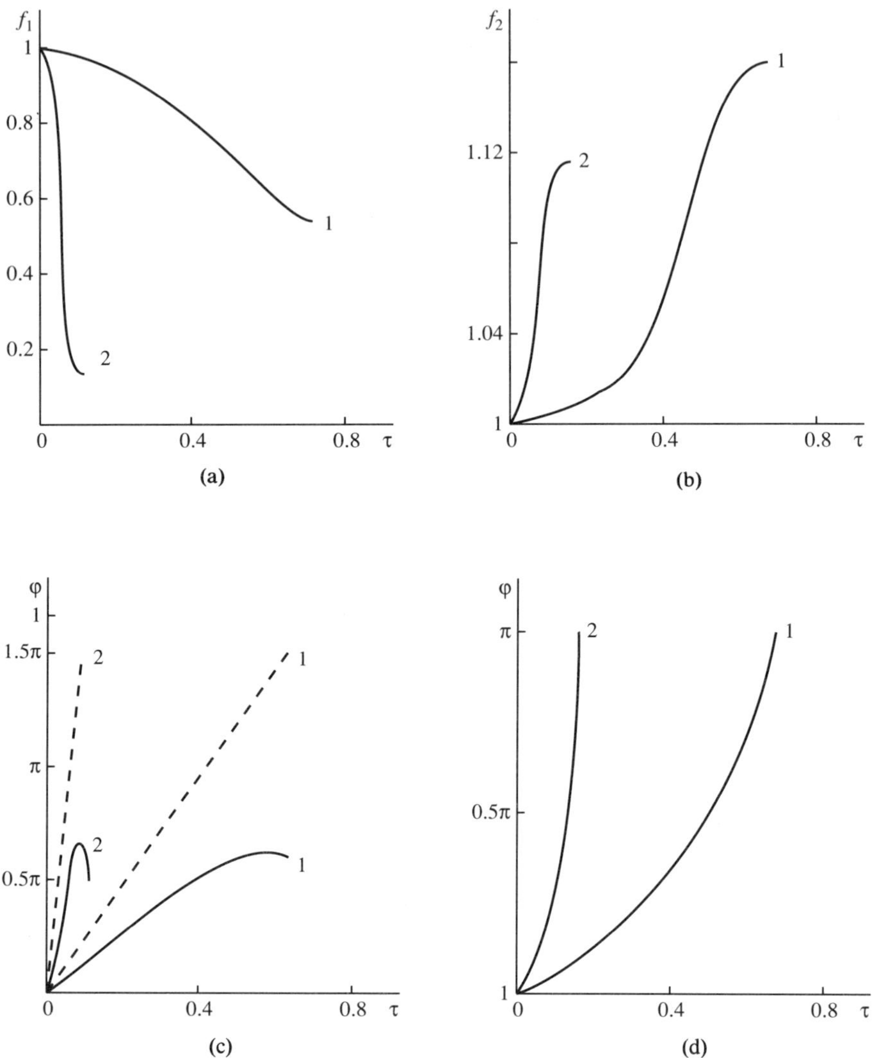

Fig. 1.19 Geometric dependence of amplitude–phase cross-modulation of the TE_{11} mode in a rectangular waveguide upon the distance τ. Curves 1 and 2 relate to the geometric parameter (1.126) of the waveguide for $u_0 = 0.5$ and $u_0 = 0.25$, respectively. (a, b, c) illustrate the modulation of amplitudes f_1, f_2 and phase shift φ between them. Curves 1 and 2 at in (d) show the modulation of phases $\varphi_{1,2}$ of each polarization component correspondingly.

The non-linear oscillations of polarization being exposed, we need to determine the periods of the oscilations. Since the square-core waveguide ($u_0 = 1$) is shown to be free from the non-linear distortions under discussion, it is instructive to present examples of solutions of the system (1.149, 1.150) related to the cases $u_0 > 1$ and $u_0 < 1$. The results of computer simulations of these problems are depicted in Fig. 1.19. Curves 1 and 2 relate to rectangular waveguides (model $\gamma = 1$) with sides $a = 1.4\,\mu\text{m}$ and $0.5\,\mu\text{m}$, respectively, while $b = 1\,\mu\text{m}$ in all these cases. The freespace wavelengths λ_0 are equal to $1.5\,\mu\text{m}$ (curve 1) and $1.06\,\mu\text{m}$ (curve 2). The spatial periods of amplitude–phase modulation and depth of modulation of amplitudes f_1 and f_2 illustrate the considerable dependence upon geometric factors. Thus the distance $z = 1\,\text{m}$ is expected to be equal to half the period of amplitude-phase modulation for these waveguides, the energy flow being about 100–150 W.

The difference in phase shifts of E_x and E_y components, shown in Fig. 1.19, indicates the non-linear splitting of the TE_{11} mode for two components propagating with slightly different phase velocities $(v_\text{p})_{1,2}$:

$$(v_\text{p}) = \frac{\omega}{\beta}\left(1 - \frac{\Delta\beta_{1,2}}{\beta}\right); \qquad \frac{\Delta\beta_{1,2}}{\beta} = \frac{\partial\varphi_{1,2}}{\partial\tau}(\beta L_\text{TE})^{-1}. \qquad (1.155)$$

The characteristic non-linear scale L_TE is given in (1.151). This difference results in complicated self-induced birefrigency connected with above-mentioned non-uniform rotation of polarization.

It is necessary to emphasize an important fact concerning this phase modulation. The sign of the phase shift far from cutoff coincides with the sign of the non-linear Kerr coefficient χ (1.88). Moreover, the phase shift $\varphi_{1,2}$ of each component of polarization far from cutoff varies monotonically. However, close to cutoff, this monotonicity may be violated for some components, as depicted in Fig. 1.20 for f_3). This graph shows that there do exist waves that change the increase of phase shift to a decrease in the course of propagation, the sign of non-linearity being the same. Hence it appears that the perturbation of propagation constant $\Delta\beta$ (1.155) may change sign during the period of self-action. The possibilities of using of this unexpected result for controlled shaping of wave pulses will be discussed in Section 3.

1.4.2 *Cross-modulation of components of the TM_{11} mode*

The analysis of cross modulation of the TM_{11} mode in a rectangular waveguide is performed in close analogy with that of TE_{11} mode. However, the three-component structure of the electric field in the TM_{11} mode involves the appearance of three equations (1.137) for self-consistent description of the non-linear evolution of this mode.

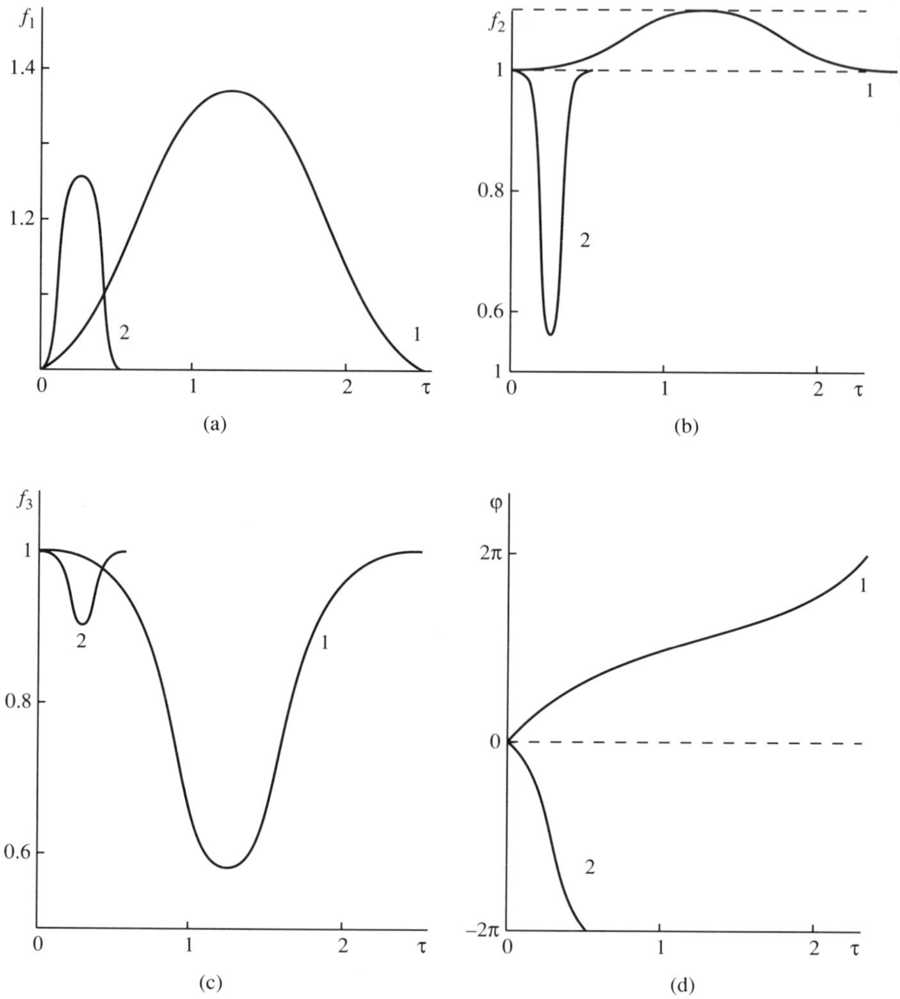

Fig. 1.20 Non-linear spatial oscillations of the TM_{11} mode polarization structure in a rectangular waveguide. Curves 1 and 2 relate to the parameters $u_0 = 4$, $y_1^2 = 0.09$, $y_2^2 = 0.36$, and $u_0 = 0.25$, $y_1^2 = 4/3$, $y_2^2 = 1/3$, respectively: (a, b, c) illustrate the oscillations of polarization components f_1, f_2, and f_3; (d) shows the self-modulation of phase shift φ between f_1 and f_2 components. τ is the dimensionless distance; the spatial periods of non-linear oscillations of amplitude–phase structure are $\tau_1 = 2.6$ and $\tau_2 = 0.4$ respectively. The phase shift φ between f_2 and f_1 components varies monotonically from 0 to $\pm 2\pi$ per one non-linear period.

Starting with the low-intensity approximation (Marcuse 1974) for the waveguide shown in Fig. 1.12,

$$E_{x0} = Ak_1 \sin k_1 x \cos k_2 y, \qquad E_{y0} = Ak_2 \cos k_1 x \sin k_2 y,$$

$$E_{z0} = +\frac{iAk_\perp^2}{\beta} \cos k_1 x \cos k_2 y, \tag{1.156}$$

let us introduce the functions

$$E_x = E_{x0} f_1 e^{i\varphi_1}; \qquad E_y = E_{y0} f_2 e^{i\varphi_2}; \qquad E_z = E_{z0} f_3 e^{i\varphi_3}; \tag{1.157}$$

$$\varphi = \varphi_2 - \varphi_1; \qquad \gamma = \varphi_3 - \varphi_1; \qquad \delta = \varphi_3 - \varphi_2. \tag{1.158}$$

The functions (1.158) determine the phase shifts between the components of the TM_{11} mode, whereby one may see that only two of these three functions are independent owing to relation $\varphi = \gamma - \delta$. Using the dimensionless variable τ, we obtain with the standard procedure the equations of self-action of the TM_{11} mode in the form:

$$\frac{\partial f_1}{\partial \tau} = -y_2^2 f_2 D_2 \sin \varphi;$$

$$\frac{\partial f_2}{\partial \tau} = y_1^2 f_1 D_1 \sin \varphi; \tag{1.159}$$

$$\frac{\partial f_3}{\partial \tau} = -\frac{T_1}{y^2};$$

$$\frac{\partial \varphi}{\partial \tau} = 8(1 + y^2)(y_2^2 f_2^2 - y_1^2 f_1^2) + y_1^2 D_1 (\cos \varphi - f_2 f_1^{-1})$$

$$\qquad - y_2^2 D_2 (\cos \varphi - f_1 f_2^{-1});$$

$$\frac{\partial \gamma}{\partial \tau} = (1 + y_2^2) D_1 - y_2^2 D_2 f_2 f_1^{-1} \cos \varphi - y^{-2} T_2 f_3^{-1}.$$

Here

$$D_1 = 2(y_2^2 f_2^2 - 3y_1^2 f_1^2 + 3y^4 f_3^2);$$

$$D_2 = 2(y_1^2 f_1^2 - 3y_2^2 f_2^2 + 3y^4 f_3^2);$$

$$T_1 = 2(f_1 y_1^2 D_1 \sin \gamma + f_2 y_2^2 D_2 \sin \delta); \tag{1.160}$$

$$T_2 = 2(f_1 y_1^2 D_1 \cos \gamma + f_2 y_2^2 D_2 \cos \delta).$$

The initial conditions are as usual:

$$f_{1,2,3} = 1; \qquad \varphi|_{\tau=0} = \gamma|_{\tau=0} = \delta|_{\tau=0} = 0; \tag{1.161}$$

$$\tau = zL_{\mathrm{TM}}^{-1}; \qquad L_{\mathrm{TM}} = \frac{16 n_0 \lambda_0 \sqrt{1 + y^2}}{\pi \chi |A|^2 k_\perp^2} . \qquad (1.162)$$

Examples of amplitude–phase cross-modulation of the TM_{11} mode shown in Figs 1.20 and 1.21 exhibit the non-harmonic non-linear oscillations of the polarization structure. The maximum deviations from polarization predicted by the low-intensity limit and described by eqn (1.161) arise at the points related to half-periods of this spatial modulation. As for the TE_{11} mode (Fig. 1.18), the maximum phase shift φ accumulated at these points owing to interaction of components of the travelling mode may result in formation of both linear ($\varphi = \pm \tau$, Fig. 1.20) and elliptical ($|\varphi| < \pi$, Fig. 1.21) polarization structures. The square-shaped waveguide ($u_0 = 1$) conserves the transversal amplitudes E_x and E_y (1.156) given by 'linear theory'; however, the modulation of phase shifts δ and γ (1.158) between longitudinal component E_z and these transversal ones does exist even in this case. The small renormalization of E_z component ΔE_z (the ratio $|\Delta E_z| \, |E_{z0}|^{-1}$ is of order of magnitude $\chi |A|^2 k_\perp^2 \ll 1$) may be calculated by analogy with (1.147).

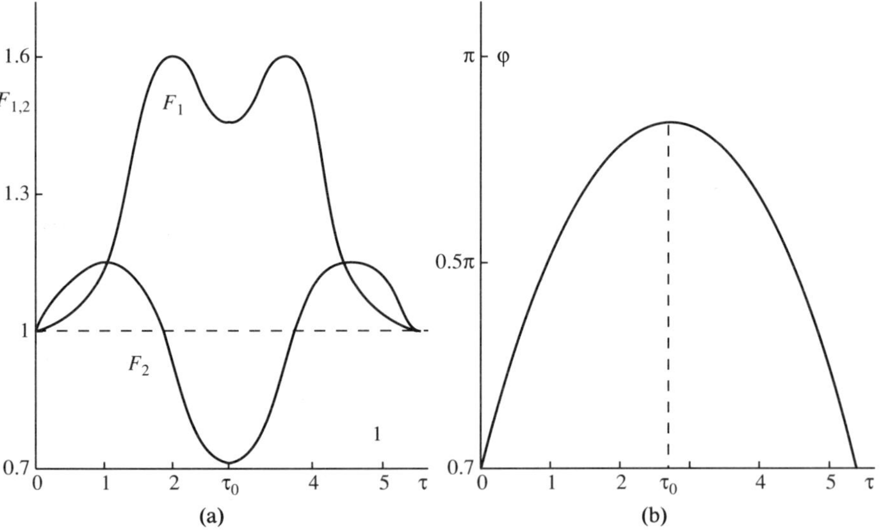

Fig. 1.21   Double-bumped waveform of spatial oscillations of the polarization structure of the TM_{11} mode in a rectangular waveguide $u_0 = 2$, $y_1^2 = 0.2$, $y_2^2 = 0.4$. (a) Two maxima for each transversal polarization component f_1 and f_2 separated by one minimum at the point $\tau_{0.5} = 0.5\tau_0$ arise inside one spatial period $\tau_0 = 5.4$. (b) Unlike Fig. 1.20 the phase shift φ between E_x and E_y components oscillates the maximum value of $|\varphi|$ being smaller than π.

Thus, the qualitatively similar tendencies of intensity-dependent evolution are observed near cutoff for both TE_{11} and TM_{11} modes. The development of these tendencies results in formation of non-linear periodical structure of the guided mode. This structure is characterized by splitting of the propagation constant of the travelling wave into two slightly different constants, modulation of amplitudes and non-linear polarization. The spatial periods of this structure are shown to depend essentially upon the waveguide geometry and the proximity of regime of propagation to cutoff. Self-modulation of other, higher TE and TM modes may be examined by means of the same scheme.

1.5 Non-linear coupling of modes in fibres

The Non-linear directional coupler is a device which permits flux-dependent transfer of energy between two adjacent optical waveguides or between two modes travelling in one waveguide. These couplers exhibit rich propagation phenomena suggestive of various optical devices such as waveguide switchers, filters, amplifiers, and all-optical gates. Coupling systems may be based on both twin-core fibres, with the cores being placed in close proximity, and one-core fibre, the physical mechanism of mode conversion being foreseen. In this the interaction of modes appears either because of the evanescent field overlap in twin-core fibres or owing to the heterogeneity of a single-core waveguide. Conversion of modes induced by some kind of heterogeneity, such as birefringency or corrugation of the core, will be analysed here. Cross-modulation of mode polarization components close to cutoff described above in Section 1.4 may be considered by analogy as a peculiar coupling of these components provided by non-local non-linear effects.

All these systems are characterized by periodic exchange of power and strongly non-linear transmission properties (Wabnitz *et al.* 1986). In view of the qualitative similarities of these power-dependent processes in different single- and twin-core couples, one may talk about the coupling of wave channels, the power distribution between channels and spatial scales of this distribution being determined by the geometry of the channel and input power. Although the exchange of power is known to occur in linear coherent couplers also (Snyder and Love 1983), the non-linear coupling essentially modifies the energetic and spatial characteristics of this exchange. The numerous designs of non-linear coherent couplers are elaborated below and we shall discuss further some examples, giving an idea of the main physical principles used in these all-optical designs.

1.5.1 *Intensity-dependent interaction of modes in twin-core couplers*

The imperfect localization of guided waves inside the cores of twin-core fibre results in formation of evanescent tails of radiation. These tails provide the variations of power in each core. The rate of such power exchange is very sensitive to the distance between the regions of effective confinement of the radiation. Since non-linearity changes the refractive indices of cores, it leads to perturbations of the scales of these regions of field confinement. This process stimulates the power-dependent variation of distance between the localized fields, the distance between the axes of the cores being invariant. Owing to this effect, the coupling of modes travelling in different cores becomes tunable.

To reveal the essential physics of much mode coupling, we may suppose the field amplitudes E_1 and E_2 in each core to be axially uniform. The spatial distribution of real mode structure and geometry of a twin-core system are characterized in this approach by some effective coupling coefficient, p. Limiting ourselves to the model of Kerr-law couplers, we may analyse the influence of both the above-mentioned linear and non-linear phenomena on the dynamics of coupling in twin-core system. To make the contribution of non-linearity essential, the phase shift between travelling modes produced by differences of propagation constants β_1 and β_2 must be comparable with the non-linear phase shift. Considering modes with slightly different constants β_1 and β_2, and generalizing the non-linear equation (1.17), we may write the pair of equations describing the interaction of modes E_1 and E_2 in the cores:

$$2i\beta \frac{\partial E_1}{\partial z} + \frac{\omega}{c} p E_2 e^{i\gamma z} + \frac{\omega^2}{c^2} \chi_1 |E_1|^2 E_1 = 0; \tag{1.163}$$

$$2i\beta \frac{\partial E_2}{\partial z} + \frac{\omega}{c} p E_1 e^{-i\gamma z} + \frac{\omega^2}{c^2} \chi_1 |E_2|^2 E_2 = 0; \tag{1.164}$$

$$\beta = \tfrac{1}{2}(\beta_1 + \beta_2); \qquad \gamma = \beta_2 - \beta_1. \tag{1.165}$$

Insight into non-linear coupling builds from the familiar theory of linear coupling, following from the set of equations (1.63, 1.64) in the limit $\chi \to 0$. Using the notation

$$\tau = \frac{z}{L_0}; \qquad L_0 = \frac{2\beta c}{p\omega}, \tag{1.166}$$

and introducing the functions $f_1(z)$, $f_2(z)$, $\theta(z)$ and the real parameter μ,

$$\theta_{1,2} = \varphi_{1,2} + \frac{\gamma z}{2},$$

$$E_1 = af_1 e^{i\theta_1}, \qquad E_2 = af_2 e^{i\theta_2}, \tag{1.167}$$

$$\theta(z) = \gamma z + \varphi, \qquad \varphi = \varphi_2 - \varphi_1,$$

we may reduce eqns (1.163, 1.164) to the following system:

$$\frac{\partial f_1}{\partial \tau} = -f_2 \sin \theta; \tag{1.168}$$

$$\frac{\partial f_2}{\partial \tau} = f_1 \sin \theta; \tag{1.169}$$

$$\frac{\partial \varphi}{\partial \tau} = \frac{f_1^2 - f_2^2}{f_1 f_2} \cos \theta. \tag{1.170}$$

Here a is the amplitude of the field E_1 at the input of the waveguide ($\tau = 0$). The solution of system (1.168–1.170) must satisfy the boundary conditions

$$f_1\big|_{\tau=0} = 1; \qquad f_2\big|_{\tau=0} = \mu; \qquad \theta\big|_{\tau=0} = \theta_0. \tag{1.171}$$

To illustrate the standard approach to solution of the equations describing the coupled modes, let us consider for simplicity the approximation $\gamma = 0$, related to identical cores. The system (1.168–1.170) now has two invariants:

$$f_1^2 + f_2^2 = 1 + \mu^2, \tag{1.172}$$

$$f_1 f_2 \cos \varphi = \mu \cos \varphi_0. \tag{1.173}$$

The solution of this system may be obtained by use of the procedure shown in Section 1.1 in eqns (1.43–1.46). Let us point out the auxillary identities following from eqn (1.68, 1.69).

$$\frac{\partial (f_1 f_2)}{\partial \tau} = F \sin \varphi, \tag{1.174}$$

$$\frac{\partial F}{\partial \tau} = -4 f_1 f_2 \sin \varphi. \tag{1.175}$$

Introduction of function F similar to the function (1.47),

$$F = f_1^2 - f_2^2, \tag{1.176}$$

proves to be useful for the series of problems under discussion.

Differentiating eqn (1.175) with respect to the variable τ and making use of invariants (1.172, 1.173), we deduce an equation governing the function F (1.176):

$$\frac{\partial^2 F}{\partial \tau^2} + 4F = 0. \tag{1.177}$$

The solution of eqn (1.177) containing the boundary conditions (1.171) is

$$f_1^2 - f_2^2 = (1 - \mu^2)\cos 2\tau - 2\mu \sin \varphi_0 \sin 2\tau. \tag{1.178}$$

Combining solution (1.178) with invariant (1.172) we obtain the distribution of power in the cores for each position τ:

$$E_1^2(\tau) = a^2(\cos^2 \tau + \mu^2 \sin^2 \tau - \mu \sin \varphi_0 \sin 2\tau), \tag{1.179}$$

$$E_2^2(\tau) = a^2(\sin^2 \tau + \mu^2 \cos^2 \tau + \mu \sin \varphi_0 \sin 2\tau). \tag{1.180}$$

The initial phase shift φ_0 in this solution is arbitrary. However, we shall discuss further the cases $\varphi_0 = 0$ and $\varphi_0 = \pi$ related to manifestation of different tendencies in interaction of modes. The last term in eqns (1.179, 1.180) containing the factor $\sin \varphi_0$ vanishes in both these cases.

According to the solution obtained, the powers in each core vary sinusoidally with the spatial period $\tau_p = \pi$. It is worth stressing an important case, when the power is injected into one core of the coupler $|E_2|_{\tau=0} = 0$. Coupling of modes will result in transfer of power to another core, the distance of total power transfer being equal to

$$\tau_0 = \frac{\pi}{2}. \tag{1.181}$$

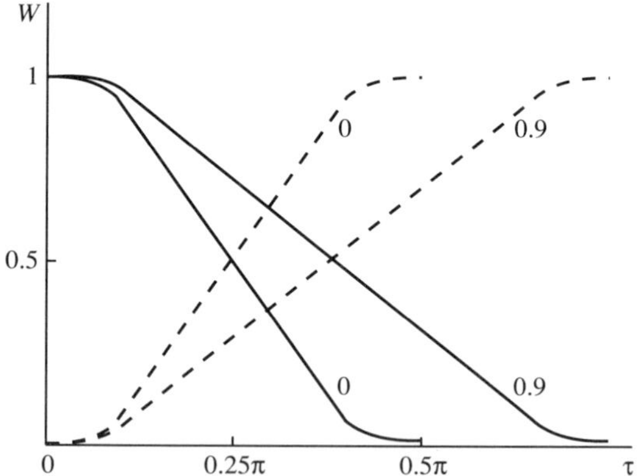

Fig. 1.22 The complete transfer of power between the cores of a twin-core waveguide due to linear ($Q = 0$) and non-linear subcritical ($Q = 0.9$) coupling of modes. The normalized intensity II_0^{-1} is plotted against the dimensionless distance τ. The solid and broken lines relate to intensities of radiation in initially excited $I|_{\tau=0} = I_0$ and initially unexcited $I|_{\tau=0} = 0$ cores, respectively.

The dynamics of distribution of powers in this low-intensity limit is shown in Fig. 1.22.

To give an idea of intensity-dependent coupling (Jensen 1982), we must generalize this scheme of analysis by reconsidering the system of equations (1.168–1.170) and reinstating the non-linear terms. It is essential that eqns (1.168, 1.169) remain invariant in the course of such reconsideration as well as identities (1.174, 1.175) derived from these equations. The generalization results only in the appearance of an additional non-linear term in eqn (1.170):

$$\frac{\partial \varphi}{\partial \tau} = \left(f_1^2 - f_2^2\right)\left(\frac{\cos \varphi}{f_1 f_2} - \frac{\chi a^2 \omega}{pc}\right). \tag{1.182}$$

Here both cores are assumed to be identical ($\beta_1 = \beta_2$; $\chi_1 = \chi_2$). The energy conservation law (1.172) is fulfilled in the non-linear problem also. Making use of eqn (1.182) we obtain the generalized form of the invariant (1.173):

$$f_1 f_2 \left(\cos \varphi - \frac{\chi a^2 \omega}{2pc} f_1 f_2\right) = \mu \left(\cos \varphi_0 - \frac{\chi a^2 \omega}{2pc}\right). \tag{1.183}$$

Now we may deduce the equation describing, instead of eqn (1.177), the same function F (1.176) connected with a non-linear coupler. Differentiating again eqn (1.175) with respect to variable τ and taking into account the identity

$$f_1^2 f_2^2 = \frac{(1 + \mu^2)^2 - F^2}{4} \tag{1.184}$$

based on invariant (1.172), we may write the generalized equation governing the function F in the form

$$\frac{\partial^2 F}{\partial \tau^2} + 4F\left[1 - \frac{\chi^2 a^4 \omega^2 (1 - \mu^2)^2}{8p^2 c^2} - \frac{\chi a^2 \omega}{pc} - \mu \cos \varphi_0\right] + \frac{\chi^2 a^4 \omega^2 F^3}{2p^2 c^2} = 0. \tag{1.185}$$

We may find the first integral of eqn (1.185) by multiplying this equation by the derivative $\partial F/\partial \tau$. To display the main tendencies in power transfer provided by phase shift φ_0, let us restrict our analysis by the cases $\varphi_0 = 0$ and $\varphi_0 = \pi$, related, according to eqn (1.175), to the vanishing of the derivative $\partial F/\partial \tau$ at the input of coupler:

$$\left.\frac{\partial F}{\partial \tau}\right|_{\tau = 0} = 0. \tag{1.186}$$

In addition, to compare the influence of coupling and non-linear phenomena on the process of power transfer, it is convenient to introduce the dimensionless parameter Q connected with the ratio of coefficients χ and p, λ_0 being the free-space wavelength:

$$Q = \frac{\pi \chi a^2}{2\lambda_0 p} \qquad (1.187)$$

The parameter Q may be either positive or negative since its sign coincides with the sign of the non-linear coefficient χ. Using condition (1.186) and notation (1.187), we obtain the first integral of eqn (1.185):

$$\frac{1}{4}\left(\frac{\partial F}{\partial \tau}\right)^2 + \left(F^2 - F_0^2\right)[1 - 2Q^2(1 - \mu^2)^2 - 4Q\mu \cos \varphi_0] + Q^2(F^4 - F_0^4) = 0. \qquad (1.188)$$

Here

$$F_0 = F|_{\tau=0} = 1 - \mu^2. \qquad (1.189)$$

Introducing the new variable T,

$$T = F^2 F_0^{-2}, \qquad T|_{\tau=0} = 1, \qquad (1.190)$$

we may rewrite eqn (1.188) in the form

$$2\tau = \frac{1}{2}\int_1^T \frac{\mathrm{d}T}{\{T(1 - T)[1 - 4Q\mu \cos \varphi_0 - Q^2(1 - \mu^2)^2(1 - T)]\}^{1/2}}. \qquad (1.191)$$

In view of condition (1.186), $\partial T/\partial \tau|_{\tau=0} = 0$; here the tendency of variation of parameter T is determined by eqn (1.185):

$$\left.\frac{\partial^2 T}{\partial \tau^2}\right|_{\tau=0} = -8(1 - 4Q\mu \cos \varphi_0). \qquad (1.192)$$

Thus, in the case $1 - 4Q\mu \cos \varphi_0 < 0$, variable T grows from the value $T = 1$ at the coupler's input to a value $T = T_0 > 1$ achieved at some point inside the coupler. The value T_0 may be calculated from eqn (1.192):

$$T_0 = 1 - \frac{1 - 4Q\mu \cos \varphi_0}{Q^2(1 - \mu^2)^2}. \qquad (1.193)$$

In the opposite case $1 - 4Q\mu \cos \varphi_0 > 1$, the value T_0 (1.193) may become either positive or negative. Here the variable T diminishes form the value $T = 1$ to the value $T_0 > 0$ or to the value $T = 0$ if $T_0 < 0$.

Basically the results (1.191–1.193) contain all the information about power distribution at any point of a directional coupler described by an inverse function $\tau = \tau(T)$ (1.191). However, to emphasize physically meaningful situations it is appropriate to discuss the problems related to injection of power into one core and into two cores separately.

A. One-core excitation ($\mu = 0$). The inverse distribution $\tau = \tau(T)$ (1.191) is reduced in this case by the substitution $T = \cos^2 x$ to an elliptic integral:

$$2\tau = \int_0^x \frac{\mathrm{d}x}{\sqrt{(1 - Q^2 \sin^2 x)}}. \tag{1.194}$$

Recalling the definition of parameter Q (1.187), one may introduce some 'critical' intensity

$$a_{\mathrm{cr}}^2 = \frac{2\lambda_0 p}{\pi |\chi|} \tag{1.195}$$

and use parameter Q proportional to the ratio of intensities

$$Q = \frac{a^2}{a_{\mathrm{cr}}^2} \operatorname{sign} \chi. \tag{1.196}$$

We may consider the solution (1.194) in three cases, classifying them with respect to the values $Q < 1$, $Q = 1$, and $Q < 1$.

With the launched intensity limited ($a^2 < a_{\mathrm{cr}}^2, Q < 1$), the function (1.194) is known to be the elliptic integral of the first kind. The function T may be expressed in terms of Jacobi elliptic functions (Ryshik and Gradstein 1957):

$$T = \operatorname{cn}^2(2\tau). \tag{1.197}$$

The definition (1.5.28) and invariant (1.172) yield the equation

$$F = f_1^2 - f_2^2 = \operatorname{cn} 2\tau. \tag{1.198}$$

From (1.198) and (1.172) it follows that

$$E_1^2 = a^2 \operatorname{cn}^2 \tau; \qquad E_2^2 = a^2 \operatorname{sn}^2 \tau \tag{1.199}$$

These solutions satisfy boundary condition (1.171), $\operatorname{sn} \tau$ and $\operatorname{cn} \tau$ being the Jacobi functions characterized by the identities (Ryshik and Gradstein 1957)

$$1 + \operatorname{cn} 2\tau = 2 \operatorname{cn}^2 \tau; \qquad 1 - \operatorname{cn} 2\tau = 2 \operatorname{sn}^2 \tau \tag{1.200}$$

The functions E_1^2 and E_2^2 (1.199) shown in Fig. 1.22 have period τ_p given by condition $\tau_p = 2K(Q)$, where $K(Q)$ is the complete elliptic integral of the first kind:

$$K(Q) = \int_0^{\frac{\pi}{2}} \frac{dx}{\sqrt{(1 - Q^2 \sin^2 x)}} .$$

(1.201)

Coupling of modes results in complete transfer of power from one core to another as well as in the linear problem discussed above. The situation under discussion $(1 \leqslant Q^2 < 1)$ is considered to be a generalization of the low-intensity limit (1.179, 1.180); the solution (1.199) is reduced to linear coupling (1.179, 1.180) when the intensity tends to zero $(Q = 0)$.

To emphasize the influence of intensity on the distance of complete intensity transfer $\tau_0 = 0.5\tau_p$, we may rewrite the value τ_0 following from (1.194) as a product of the value $\pi/2$ related to the linear case and a limited dimensionless factor $S(Q)$:

$$\tau_0 = \frac{\pi}{2} S(Q).$$

(1.202)

This factor $S(Q) = 2\pi^{-1} K(Q)$ is equal to 1 in the low-intensity limit $(Q = 0)$ and grows monotonically when $Q > 0$.

The opposite situation $(Q^2 > 1)$ related to intensity a^2 exceeding the critical value a_{cr}^2 may be treated in a similar fashion. Introducing in (1.194) the new variable u to according to the condition

$$Q \sin x = \sin u,$$

(1.203)

we may find from (1.194)

$$2\tau = \frac{1}{Q} \int_0^u \frac{du}{\sqrt{(1 - Q^{-2} \sin^2 u)}} .$$

(1.204)

Integration in (1.204) yields

$$\sin x = \frac{\mathrm{sn}(2\tau Q)}{Q} .$$

(1.205)

Calculating the function $T = \cos^2 x$, we find the intensities in both cores:

$$E_{1,2}^2 = \frac{a^2}{2} \left[1 \pm \sqrt{\left(1 - \frac{\mathrm{sn}^2(2\tau Q)}{Q^2}\right)^{1/2}} \right].$$

(1.206)

The coupling of modes in this case $(Q > 1)$ results in incomplete power transfer (Fig. 1.23). The distance of maximal power transfer is expressed by means of an elliptic integral (1.201)

$$\tau_0 = \frac{1}{2Q} K\left(\frac{1}{Q}\right).$$

(1.207)

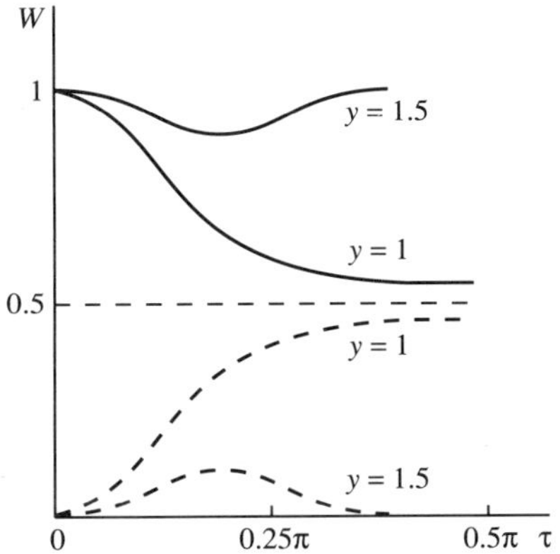

Fig. 1.23 The incomplete transfer of power of co-propagating modes between the cores of a double-core coupler in the critical ($Q = 1$) and supercritical ($Q = 1.5$) regimes of mode coupling, the power being launched into one core only. The solid and broken lines correspond to dimensionless intensities of waves W in initially excited and unexcited cores.

If the intensity of injected wave is equal to the 'critical' value (1.195) ($Q = 1$), the non-linear evolution leads, unlike the above-mentioned periodical power transfer, to asymptotic formation of equal powers in both cores:

$$E_{1,2} = \frac{a^2}{2}\left(1 \pm \frac{1}{\text{ch } 2\tau}\right) \tag{1.208}$$

To explain these figures related to initial power injection to one core of twin coupler it is worth comparing them with some results in consideration of couplers with both cores being excited.

B. Coupling of modes excited in both cores ($\mu > 0$) The diversity of results of interaction of modes launched in both channels of twin-core fibre is enriched by the influence of three additional free parameters of the problem, which vanish in the case of single-core excitation:

1. the ratio of intensities given by parameter μ;
2. the phase shift φ_0;
3. the sign of non-linearity (positive or negative) determining the sign of constant Q (1.196).

To illustrate the influence of these parameters we discuss the cases $T_0 = 1$ and $T_0 = 0$ in eqn (1.193). Both powers and phases in each channel are conserved in the case $T = 1$, i.e.

$$4\mu Q \cos \varphi_0 = 1. \tag{1.209}$$

This new stationary distribution may arise only due to excitation of both cores ($\mu > 0$). The condition (1.209) being fulfilled, eqn (1.192) yields that $\partial^2 \varphi / \partial \tau^2 = 0$ as well as $\partial \varphi / \partial \tau = 0$ if $\varphi_0 = 0$ or $\varphi_0 = \pm \pi$. It is important to stress that interaction between modes may vanish according to (1.209) both in focusing ($Q > 0$, $\varphi_0 = 0$) and defocusing ($Q < 0$, $\varphi_0 = \pi$) media.

Another peculiarity is determined by the condition $T_0 = 0$, which results in.

$$1 - 4\mu Q \cos \varphi_0 = Q^2 (1 - \mu^2)^2. \tag{1.210}$$

Integration of (1.193) shows in this case that the power tends to be distributed asymptotically in equal parts in each core:

$$E_{1,2}^2 = \frac{a^2}{2} \left[1 + \mu^2 \pm \frac{1 - \mu^2}{\operatorname{ch} 2\tau} \right]. \tag{1.211}$$

Other cases related to arbitrary values of parameter T_0 (1.193) in eqn (1.191) are reduced to standard elliptic integrals of the first kind.

It is necessary to stress that integration in eqn (1.193) is needed only for calculation of distance the τ_0 related to formation of maximum power in one core, and respectively the minimum power in the other core, while the powers E_1^2 and E_2^2 in position τ_0 may be determined without integration. Since parameter T (1.190) describing the evolution of coupled modes takes the value $T = T_0$ at the point $\tau = \tau_0$, we may obtain these powers from eqns (1.172) and (1.190) with $T = T_0 \geqslant 0$:

$$E_{1,2}^2(\tau_0) = \frac{a^2}{2} \left[1 + \mu^2 \pm (1 - \mu^2)\sqrt{T_0} \right]. \tag{1.212}$$

This formula contains all the information about the power levels achieved due to coupling of modes.

Reviewing these results for all-optical couplers we may outline the main features determining the competition of linear effects, such as tunnel penetration of modes between cores, and non-linear self-action of guided fields. The non-linearity slows the rates of power exchange between cores provided by linear effects; however, this tendency dominates, only one channel being excited and the power launched into this channel being limited, $a^2 < a_n^2$ (Fig. 1.22). The proximity of the intensity to its 'critical' value ($Q^2 \to 1$) results in considerable increase in the period of this exchange. The increase of power launched into one core ($Q \to 1$) restricts the power transfer between cores owing to self-confinement of energy flows (Fig. 1.23). Simultaneous excitation of two channels of the coupler provides the rich variety of spatial

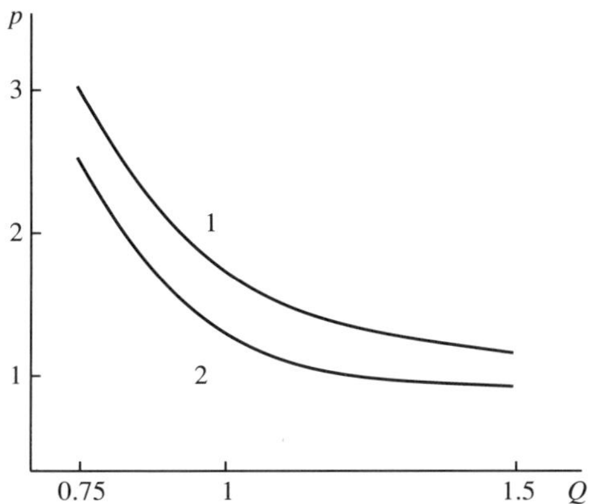

Fig. 1.24 Spatial periods of non-linear oscillative transfer of power between cores in a twin-core waveguide, both cores being excited at the waveguide's input. The periods τ_0 are plotted against non-linear parameter Q (1.196). Curves 1 and 2 relate to the values $\mu = 0.2$ and $\mu = 0.5$, respectively, $\varphi_0 = 0$.

periods of coupling τ_0 compared with the 'linear' value $\tau_0 = \pi/2$ shown in Fig. 1.24. It is useful to note that the change of sign of non-linearity ($\chi \rightarrow -\chi$) accompanied by the change of phase $\varphi_0 \rightarrow \pi - \varphi_0$, other parameters being the same, leaves invariant the picture of mode coupling, determined by eqn (1.191).

The energetic characteristics and spatial scales of such couplers are given, for example, by eqns (1.202) and (1.203). The values $\chi = 5 \times 10^{-19}$ SI (InSb), power about $P = 10$ W, cross section of waveguide $S = 3 \times 10^{-8}$ cm^2, $p = 10^{-4}$ cm^{-1}, $n_0 = 3.6$, and $\lambda_0 = 1.5$ μm give a length of power transfer of about $L_0 = 1$ cm (Jensen 1982). A similar evaluation for glass fibre ($\chi = 10^{-22}$ SI) predicts a length of $L_0 = 20$ cm.

These examples of interaction between tunnel penetration and non-linear self-action of guided modes illustrate the physical foundations of various types of twin-core mode couplers. The mathematical approach to these problems is based on use of elliptical functions. The results obtained in subsection B are shown to reduce to the related results of subsection A in the limit $\mu = 0$, as well as the results of subsection A in their turn tending to corresponding 'linear' formulae in the low-intensity limit ($Q \rightarrow 0$). These detailed descriptions will be used further as examples of the solution of numerous problems in non-linear integrated optics connected with self-action of CW guided modes.

1.5.2 *Non-linear tunable polarizers and sensors based on single-core birefringent and gyrotropic waveguides*

Birefringent optical fibres are interesting for the formation of large intensity-dependent phase shifts between orthogonally polarized modes. The fibres' non-linearity is expected to produce an essential influence on mode coupling, the spatial scales of both linear mode conversion and non-linear phase self-modulation being of the same order of magnitude. It appears that the small difference between refractive indices of interacting modes will be comparable in this case with the non-linear perturbations of these indices. Alongside power-dependent gyrotropy of initially isotropic non-linear media (see Section 1.1), some small birefringence independent of wave power may be inherent to the fibre owing to the technology of fabrication. The competition between these linear and non-linear processes in the coupling of interacting modes may provide the formation of new polarization states in guided radiation. Unlike the components of sensor systems based on inelastic non-linear processes, such as stimulated Raman and Brillouin scattering, elastic processes conserving the wave frequency are shown to be useful for elaboration of fibre sensors too.

To analyse these possibilities, let us consider the interaction of two orthogonally polarized waves $\mathbf{E}_1$ and $\mathbf{E}_2$, launched into a slightly birefringent fibre (Winful 1985), supposing for simplicity that the directions of vectors $\mathbf{E}_1$ and $\mathbf{E}_2$ coincide with the principal polarization axes (x, y) in fibre's cross section. The linear dielectric tensor $\mathscr{E}_{ij}$ is assumed to be diagonal in the principal axes representation:

$$\mathscr{E}_{ij} = \mathscr{E}_0 \begin{pmatrix} \mathscr{E}_{\perp 1} & 0 \\ 0 & \mathscr{E}_{\perp 2} \end{pmatrix} \tag{1.213}$$

Denoting the propagating constants related to these axes as β_1 and β_2, keeping in mind the smallness of the birefringence ($|\beta_2 - \beta_1| \ll \beta_2 + \beta_1$) and using formula (1.6) for the non-linear susceptibility of the waveguide medium, we may write the equations describing the interaction of these waves in the form

$$2i\beta \frac{\partial \mathbf{E}_1}{\partial z} + p \frac{\omega}{c} \mathbf{E}_2 + \frac{\omega^2}{c^2} \chi [\,(\,|\mathbf{E}_1|^2 + A\,|\mathbf{E}_2|^2)\,\mathbf{E}_1 + B\mathbf{E}_2^2\mathbf{E}_1^*] = 0;$$

$$\tag{1.214}$$

$$2i\beta \frac{\partial \mathbf{E}_2}{\partial z} + p \frac{\omega}{c} \mathbf{E}_1 + \frac{\omega^2}{c^2} \chi [\,(\,|\mathbf{E}_2|^2 + A\,|\mathbf{E}_1|^2)\,\mathbf{E}_2 + B\mathbf{E}_1^2\mathbf{E}_2^*] = 0.$$

Here p is the coupling constant, $\beta = \tfrac{1}{2}(\beta_1 + \beta_2)$. Introducing the standard notations

$$E_1 = af_1(z)e^{i\theta_1(z)}, \qquad E_2 = af_2(z)e^{i\theta_2(z)},$$

$$\theta_{1,2}(z) = \varphi_{1,2}(z) \mp \frac{\gamma z}{2}, \qquad \gamma = \beta_2 - \beta_1, \tag{1.215}$$

where the phase shifts $\theta_{1,2}$ include both intensity-independent $(\pm\gamma/2)$ and intensity-dependent $(\varphi_{1,2})$ parts, we may transform eqn (1.214) to a dimensionless system convenient for analysis:

$$\frac{\partial f_1}{\partial \tau} = -f_2 \sin\theta (1 + Qf_1 f_2 \cos\theta);$$

$$\frac{\partial f_2}{\partial \tau} = f_1 \sin\theta (1 + Qf_1 f_2 \cos\theta); \tag{1.216}$$

$$\frac{\partial \theta}{\partial \tau} = (f_2^2 - f_2^2)\left[\frac{\cos\theta}{f_1 f_2}(1 + Qf_1 f_2 \cos\theta) - Q\right]. \tag{1.217}$$

Here

$$\theta = \theta_1 - \theta_2; \qquad Q = \frac{2\chi Ba^2}{p}. \tag{1.218}$$

The role of non-linearity is characterized by the parameter Q (1.218), which may be either positive or negative depending on the sign of the Kerr coefficient χ. The normalized distance τ is determined by analogy with (1.166):

$$\tau = \frac{z}{L_0}; \qquad L_0 = \frac{2\beta c}{p\omega}. \tag{1.219}$$

The boundary conditions to eqns (1.215–1.217) are

$$f_1|_{\tau=0} = 1; \qquad f_2|_{\tau=0} = \mu; \qquad \theta|_{\tau=0} = \theta_0. \tag{1.220}$$

The exact analytical solution of these equations may be obtained by means of two conservation laws, the value of non-linear parameter Q being arbitrary:

$$f_1^2 + f_2^2 = 1 + \mu^2, \tag{1.221}$$

$$2f_1 f_2 \cos\theta - Q(f_1 f_2 \sin\theta)^2 = 2\mu \cos\theta_0 - Q\mu^2 \sin^2\theta_0. \tag{1.222}$$

Invariant (1.222) may be considered as a generalization of two previously analysed special problems related to the cases $Q \to 0$ (low-intensity limit) and $|Q| \gg 1$ (high-intensity limit). In the first case $(Q \to 0)$ invariant (1.222) obviously reduces to (1.173) corresponding to the linear coupling of modes. The limit $|Q| \gg 1$ leads to transformation of eqn (1.222) to (1.42) related to non-linear interaction of two orthogonal components of polarization, the linear coupling between these components being ignored. Generalizing the approach advanced above (Sections 1.1. and 1.5.2), let us introduce the functions F and R:

$$F = f_1^2 - f_2^2; \qquad R = 1 + Qf_1f_2\cos\theta. \tag{1.223}$$

These functions obey auxiliary equations derived from the system (1.216–1.217):

$$\frac{\partial F}{\partial \tau} = -4f_1f_2R\sin\theta, \tag{1.224}$$

$$\frac{\partial(f_1f_2)}{\partial \tau} = FR\sin\theta. \tag{1.225}$$

Combining eqns (1.224, 1.225) with eqn (1.217) and (1.222) we obtain

$$\frac{\partial^2 F}{\partial \tau^2} + 4F[1 + 3Qf_1f_2\cos\theta - 2\mu\cos\theta_0 + Q^2\mu^2\sin^2\theta_0] = 0. \tag{1.226}$$

To express the product $Qf_1f_2\cos\theta$ in terms of the function F we may use the conservation law (1.222) and the identity (1.184):

$$Qf_1f_2\cos\theta = -1 + \frac{Q}{2}\sqrt{D^2 - F^2}, \tag{1.227}$$

$$D^2 = \frac{4(1 + Q\mu\cos\theta_0)^2}{Q^2} + (1 - \mu^2)^2. \tag{1.228}$$

Substitution of (1.227) into (1.226) yields the equation governing the function $F(\tau)$:

$$\tfrac{1}{4}\frac{\partial^2 F}{\partial \tau^2} + F[-2(1 + Q\mu\cos\theta_0) + Q^2\mu^2\sin^2\theta_0 + \tfrac{3}{2}Q\sqrt{D^2 - F^2}] = 0. \tag{1.229}$$

The function F satisfies boundary conditions

$$F|_{\tau=0} = 1 - \mu^2; \qquad \left.\frac{\partial F}{\partial \tau}\right|_{\tau=0} = -4\mu(1 + Q\mu\cos\theta_0)\sin\theta_0. \tag{1.230}$$

The first integral of the non-linear equation (1.229) may be written as

$$\frac{1}{4}\left(\frac{\partial F}{\partial \tau}\right)^2 = Q(D^2 - F^2)\left[\sqrt{D^2 - F^2} - \frac{2(1 + Q\mu\cos\theta_0)}{Q} + Q\mu^2\sin^2\theta_0\right]. \tag{1.231}$$

Using the substitution

$$F = D \sin \alpha \tag{1.232}$$

we may derive from eqn (1.231)

$$2\tau\sqrt{D|Q|} = \int_{\alpha_0}^{\alpha} \frac{d\alpha}{\sqrt{\cos\alpha - \dfrac{2(1 + Q\mu\cos\theta_0)}{DQ} + \dfrac{Q\mu^2\sin^2\theta_0}{D}}} , \tag{1.233}$$

$$\alpha_0 = \arcsin\left(\frac{1 - \mu^2}{D}\right). \tag{1.234}$$

Integral (1.233) may be expressed in terms of elliptic functions. To complete the solution using the function $F(\tau)$ (1.233), we must calculate the unknown functions f_1, f_2 and θ by means of eqn (1.221):

$$f_{1,2}^2 = \frac{1 + \mu^2 \pm F}{2} , \tag{1.235}$$

$$\cos\theta = \frac{\sqrt{D^2 - F^2} - 2Q^{-1}}{\sqrt{[(1 + \mu^2)^2 - F^2]^{1/2}}} \tag{1.236}$$

Although the solution obtained (1.233) contains all the information about the evolution of modes $\mathbf{E}_1$ and $\mathbf{E}_2$, the straightforward analysis of this general solution is impeded by the simultaneous influence of various parameters, such as p, Q, μ, and θ_0. The peculiarities of mode evolution in the problem under discussion may be illustrated by means of some important special cases related to limited ranges of variation of parameter χa^2 or fixed values of phase $\theta_0 (\theta_0 = 0, \pm\pi/2)$. These cases will be analysed further for birefringent and gyrotropic waveguides separately.

A. Tunable birefringent polarizer Let us consider the influence of initial phase shift θ_0 on the dynamics of mode interaction using the examples of linear and circular polarization and the manifestations of different tendencies in such interaction.

(a) Linear polarization ($\theta = 0$). This case, relating to the splitting of a linearly polarized wave $\mathbf{E}$ incident up on the fibre's input into two waves $\mathbf{E}_1$ and $\mathbf{E}_2$ travelling along the fibre, is characterized by conditions (1.224) and (1.226).

$$\left.\frac{\partial F}{\partial\tau}\right|_{\tau=0} = 0; \qquad \left.\frac{\partial^2 F}{\partial\tau^2}\right|_{\tau=0} = -4(1 - \mu^2)(1 + Q\mu). \tag{1.237}$$

Supposing, for definitness, $\mu < 1$ one can see that the derivative $\partial^2 F/\partial \tau^2$ is negative for the condition $1 + Q\mu > 0$. Hence it follows from eqn (1.231) that the function F at the beginning of evolution diminishes to the value $F_1 = -F_0$; subsequently the function F grows to the value $F_2 = F_0$, and so on. The quarter of the spatial period of these oscillations related to the value $F = 0$ may be calculated from eqn (1.233):

$$\tau_0 = \frac{K(q)}{\sqrt{\left[(1 + Q\mu)^2 + \dfrac{Q^2}{4}(1 - \mu^2)^2\right]}}, \tag{1.238}$$

$$q = \frac{1}{\sqrt{2}}\sqrt{\left(1 - \frac{2(1 + Q\mu)}{QD}\right)}.$$

Here $K(q)$ is the complete elliptic integral of the first kind. The state of polarization at the point $\tau = \tau_0$ is characterized by the values $f_1 = f_2$ and $\theta = \theta_{\mathrm{m}}$:

$$f_1 = f_2 = \sqrt{\left(\frac{1 + \mu^2}{2}\right)}; \qquad \theta_{\mathrm{m}} = \arccos\left(\frac{D - 2Q^{-1}}{1 + \mu^2}\right). \tag{1.239}$$

It is useful to mention that in the case of linear polarization ($\theta_0 = 0$) parameter μ is connected with the angle δ between x-axis of polarization and the direction of the electric field at the waveguide input:

$$\mu = \mathrm{tg}\,\delta. \tag{1.240}$$

Thus the dependence of spatial scale τ_0 (1.238) upon the parameters μ and Q illustrates power-dependent polarization in a slightly anisotropic medium (1.213). The low-intensity limit of eqn (1.238) is

$$\lim \tau_0\big|_{Q=0} = \frac{\pi}{2}, \tag{1.241}$$

obtained in eqn (1.181) for the linear problem.

(b) Circular polarization ($\theta_0 = \pm\pi/2$, $\mu = 1$). Although the solution (1.233) is valid, with the values of parameter μ being arbitrary, it is worth emphasizing here the case $\mu = 1$, related to circularly polarized waves, because the non-linear polarization of such waves is manifested only due to interaction of both linear and non-linear coupling. The weakening of linear coupling ($p \to 0$) will be shown to provide the invariance of circular polarization state in a non-linear Kerr-law medium. Thus, considering the case ($\theta_0 = \pi/2$, $\mu = 1$, right circular

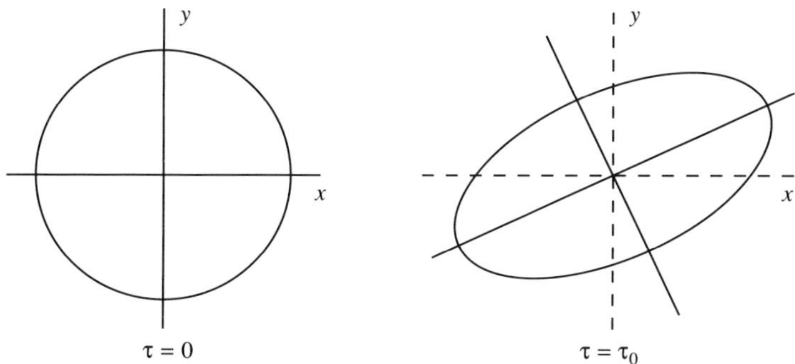

Fig. 1.25 Tunable spatial oscillations of wave polarization in a non-linear ($Q = 4$ (1.218)) birefringent polarizer from circular ($\tau = 0$) to elliptical ($\tau = \tau_0$) polarization states and vice versa. The orientation of the ellipse's major axis also oscillates.

component), we obtain from eqn (1.233) in the case of high wave intensity $|Q| \gg 2$

$$F = -\frac{2}{|Q|}\,\mathrm{sn}\,(\tau|Q|).\tag{1.242}$$

The self-action results in periodic transformation of circular polarization to elliptical and vice versa (Fig. 1.25). This ellipse of polarization is determined by functions $f_{1,2}^2$, the phase shift θ_m between oscillations in the x and y directions, and the inclination of its major axis. The functions f_1^2 and f_2^2 achieve their minimum and maximum values, respectively,

$$f_{1,2}^2 = 1 \pm \frac{1}{|Q|}\tag{1.243}$$

at the point

$$\tau_0 = \frac{1}{|Q|}\,K\!\left(\frac{2}{|Q|}\right).\tag{1.244}$$

The phase shift θ_m at the point τ_0

$$\theta_\mathrm{m} = \arccos\left[-\frac{1}{\sqrt{Q^2 - 1}}\right]\tag{1.245}$$

is twice the angle δ_m between the major axes of the polarization ellipse and x-axis, here $\delta_\mathrm{m} < 0$. The initial circular polarization is reinstated at the point $\tau = 2\tau_0$.

In the opposite case ($\theta_0 = -\pi/2$, $\mu = 1$, left circular component), the polarization state at the same point $\tau = \tau_0$ (1.244) is characterized by the values

$$f_{1,2}^2 = 1 \pm \frac{1}{|Q|}; \qquad \delta = \delta_{\mathrm{m}} > 0.$$

The growth of the non-linear parameter $|Q|$ ($|Q|) \gg 2$) leads to the diminishing of variations of both functions f_1^2 and f_2^2 ($\lim f_{1,2}^2|_{Q \to \infty} = 1$), i.e. to weakening of linear coupling between E_x and E_y polarizations.

B. Tunable gyrotropic polarizer   Before discussing the high-intensity limit in the dynamics of non-linear polarization in an anisotropic medium with diagonal dielectric tensor $\mathscr{E}_{ij}$ (1.213), it will be useful to consider the problem in a slightly gyrotropic medium, described in a linear approximation by the non-diagonal Hermitian dielectric tensor $\mathscr{E}_{ij}$:

$$\mathscr{E}_{ij} = \mathscr{E}_0 \begin{pmatrix} \mathscr{E}_\perp & ig \\ -ig & \mathscr{E}_\perp \end{pmatrix}. \tag{1.246}$$

It is interesting to analyse here the small non-diagonal elements ($|g| \ll \mathscr{E}_\perp$) which may become comparable with non-linear perturbations of refractive index n_0 determined by diagonal elements $n_0 = \sqrt{\mathscr{E}_\perp}$. Such a case seems to be very real, e.g. in elaboration of magneto-optical fibre sensors for measurement of external magnetic fields by means of the Verdet effect, inducing the small gyrotropy of an initially isotropic medium. Let us consider the elliptically polarized wave, excited near the fibre's input; the principal axes of this ellipse are E_x and E_y (Fig. 1.1). In describing the non-linear evolution of such a wave we can convince ourselves that this problem has a close relationship to the analogous problem of an anisotropic medium. Starting with equations similar to (1.214):

$$2i\beta \frac{\partial E_x}{\partial z} + i\frac{\omega^2}{c^2} gE_y + \frac{\omega^2}{c^2} \chi[(E_x(|E_x|^2 + A|E_y|^2) + BE_y^2 E_x^*] = 0,$$

$$\tag{1.247}$$

$$2i\beta \frac{\partial E_y}{\partial z} - i\frac{\omega^2}{c^2} gE_x + \frac{\omega^2}{c^2} \chi[(E_y(|E_y|^2 + A|E_x|^2) + BE_x^2 E_y^*] = 0,$$

and using the notation (1.215, 1.220), we obtain a system of equations for non-linear gyrotropic wave phenomena:

$$\frac{\partial f_1}{\partial \tau} = -Tf_2 \cos\theta; \tag{1.248}$$

$$\frac{\partial f_2}{\partial \tau} = Tf_1 \cos \theta; \tag{1.249}$$

$$\frac{\partial \theta}{\partial \tau} = -\frac{T(f_1^2 - f_2^2) \sin \theta}{f_1 f_2} . \tag{1.250}$$

Here

$$T = 1 + Qf_1 f_2 \sin \theta; \qquad Q = \frac{2\chi Ba^2}{g} ; \tag{1.251}$$

$$\tau = z L_0^{-1}; \qquad L_0 = \frac{2c^2 \beta}{\omega^2 g} . \tag{1.252}$$

Using two conservation laws of this system coinciding with (1.41) and (1.42) we may obtain the equation governing the function $F = f_1^2 - f_2^2$:

$$\frac{\partial^2 F}{\partial \tau^2} + 4FT^2 = 0. \tag{1.253}$$

Owing to invariance of the product $f_1 f_2 \sin \theta$ following from eqn (1.42), parameter T (1.251) remains constant,

$$T = 1 + Q\mu \sin \theta_0. \tag{1.254}$$

In consequence of conservation of parameter T, eqn (1.253) proves to be a linear equation; choosing the initial phase shift $\theta_0 = \pm\pi/2$ we can write the solution of eqn (1.253) in the form

$$F = (1 - \mu^2) \cos (2T\tau). \tag{1.255}$$

Following the description of power-dependent rotation of polarization ellipse, given in Section 1.1, we may conclude that this ellipse in the non-linear gyrotropic medium turns with uniform angular speed, the ellipse's shape being conserved (Fig. 1.1). The angle between the current orientation of the major axis and its initial orientation will reach the value $\pi/2$ at the point

$$\tau_0 = \frac{\pi}{2T}; \qquad z_0 = L_0 \tau_0 = \frac{\pi \beta c^2}{\omega^2 g T} . \tag{1.256}$$

The distance z_0 (1.256) relates to half of the period of non-linear rotation of the polarization; the parameter T is given in (1.254).

Generalizing the picture of power-dependent turning of the polarization ellipse in an isotropic medium (Section 1.1) we may derive from eqn (1.250) that the conclusion about the opposite directions of this ellipse turning and electric vector rotation (given in Section 1.1) is valid as long as the value of the parameter T is positive. Here the cases $Q \sin \theta_0 > 1$ and $Q \sin \theta_0 < 1$ relate to the non-linear amplification or

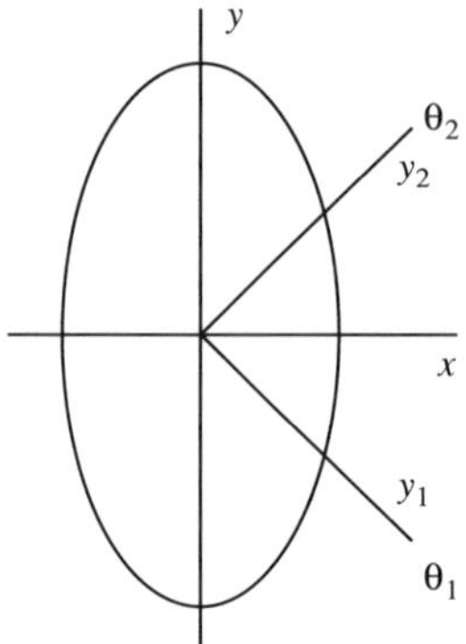

Fig. 1.26 Non-linear amplification and weakening of the turning angle θ of the polarization ellipse in the cases T > 1 and $T < 1$ (1.254), respectively, due to interaction of linear and non-linear gyrotropic effects in the waveguide. The dimensionless distance of evolution is assumed to be constant $\tau = \pi/2$ ($T = 1$), $\theta_{1,2} = \theta$ ($T_{1,2}$).

weakening of turning of the polarization ellipse in a gyrotropic medium well known from traditional linear theory ($Q = 0$, $T = 1$) (Fig. 126).

The non-linear stabilization of initial polarization structure in a gyrotropic medium occurs for the case $T = 0$,

$$Q\mu \sin \theta_0 = -1. \tag{1.257}$$

The interaction of linear and non-linear gyrotropic effets in the case $T < 0$ results according to eqn (1.250) in a change of the direction of the ellipse's turning to the opposite.

Thus, slightly anisotropic and gyrotropic single-core fibres may be used as power-dependent tunable distributed polarizers for guided waves. On the other hand, the modulation of polarization at the input of waveguide system can provide the intensity modulation at the system's output. This effect seems to be attractive for elaboration of non-linear fibre sensors.

C. Sensor-related applications Sensor-related application of these phenomena are based on the possibility of non-linear amplification of phase shifts in the polarization structure of a wave travelling in a fibre. These shifts may be produced by perturbations of a fibre acting as a sensor. Thus the presence of external magnetic field H is known to provide the turning of the electric vector **E** of a linearly polarized wave. The azimuth δ determining the wave polarization at some point z is

$$\delta = VHz. \tag{1.258}$$

Here V is the Verdet coefficient. Comparing eqn (1.258) with the well-known expression for the turning angle δ_0 in the linear gyrotropic medium,

$$\delta_0 = \frac{\omega}{2c} \frac{g}{n_0} z, \tag{1.259}$$

we can find the components of the dielectric tensor $\mathscr{E}_{ij}$ (1.246) connected with the forced gyrotropy,

$$g = 2cn_0\omega^{-1}VH. \tag{1.260}$$

Substituting g from (1.260) into the definitions of parameters τ and Q (1.251, 1.252), we may use the advanced theory of power-dependent rotation of the polarization ellipse for choosing a suitable regime for non-linear tuning of this rotation. Thus, fixing the direction of the major axis of the polarization ellipse at the waveguide's input ($x-$axis), one can conclude from eqn (1.256) that the angle of inclination of this axis in the course of the ellipse's turning will achieve the value $\delta = \pi/2$ ($\theta_0 = -\pi/2$) at the point $\tau_0 = \pi/2$ related in the low-intensity limit ($T = 1$) to the waveguide's length, $z_0 = (\pi/2) L_0$. However, the non-linear effect characterized, for example, by the value $T = 0.5$, leads according to eqn (1.256) to a decrease of this inclination angle at the waveguide's output down to the value $\delta = \pi/4$. The analyser located at the waveguide's output provides filtration of the polarized signal. Supposing the transmission plane of the analyser to be parallel to the major axis of the polarization ellipse at the waveguide's input, we may describe the tunable transmittivity of this system, characterized by the ratio of intensities $W(z_0)$ and W_0 at the input and output of waveguide, respectively:

$$K = \frac{W(z_0)}{W_0} = \frac{\cos^2\delta + \mu^2\sin^2\delta}{1 + \mu^2}. \tag{1.261}$$

The transmission coefficient K depends essentially upon the intensity of the travelling wave: thus, the use of the above-mentioned examples $\delta|_{T=1} = \pi/2$ and $\delta|_{T=0.5} = \pi/4$ leads to the values $K|_{T=1} = \mu^2(1 + \mu^2)^{-1}$; $K|_{T=0.5} = 0.5$. Assuming, for example, that $\mu = 0.5$, $\theta_0 = -\pi/2$, we can find the related non-linear parameters $Q|_{T=1} = 0$, $Q|_{T=0.5} = 1$ and calculate the ratio $K_1K_0^{-1} = 2.5$. The factor of importance is the intensity-dependent growth of transmittivity illustrated by this ratio. To evaluate the spatial scales and energy flows corresponding to these dimensionless parameters, let us consider, for example, the propagation of radition with free-space wavelength $\lambda_0 = 1.5\,\mu m$

along a silica fibre ($n_0 = 1.5$) characterized by Verdet coefficient $V = 5 \times 10^{-4}\,\mathrm{rad\,T^{-1}\,m^{-1}}$ and located in a constant magnetic field $H = 10^{-2}\,\mathrm{T}$. Calculations with these parameters using eqn (1.252) predict the values $z_0 = 30\,\mathrm{m}$, $P = 10^8\,\mathrm{W\,cm^{-2}}$, which provide the above-mentioned non-linear increase of the fibre's transmittivity by 2.5 times. This variation of energy flow may be utilized for elaboration of magneto-optical sensors.

The device under consideration is based on the non-linear properties of the sensing fibre itself. An another sensor-aided usage of power-dependent gyrotropy is connected with separation of sensing and non-linear waveguides. A schematic of a set-up using these separated waveguides is shown in Fig. 1.27. A pump laser beam propagates along two independent single-mode fibres; the upper one is the sensing fibre and the lower one is the reference fibre. The polarization system P located at the input of these fibres provides the polarization of the beams trapped in the sensing and reference fibres in orthogonal directions X and Y. When the sensing fibre is perturbed by temperature or stress, this results in variation of the refractive index and, consequently, to a phase shift φ between the waves travelling along perturbed and unperturbed fibres. The interference of waves passing through these fibres results in the formation of an elliptically polarized wave with phase shift φ between the components. Subsequently this ellipse of polarization turns owing to non-linear self-action during propagation along the capillary C filled with a liquid with considerable non-linear susceptibility (e.g. CS_2), described by the model (1.6). The analyser A, situated at the output of the capillary, provides filtration of the non-linearly transformed signal. Thus the result of this transformation depends upon the perturbation of the sensing fibre.

The dependence of the evolution of the elliptically polarized wave formed at the waveguide's input upon the phase shift θ_0 between the

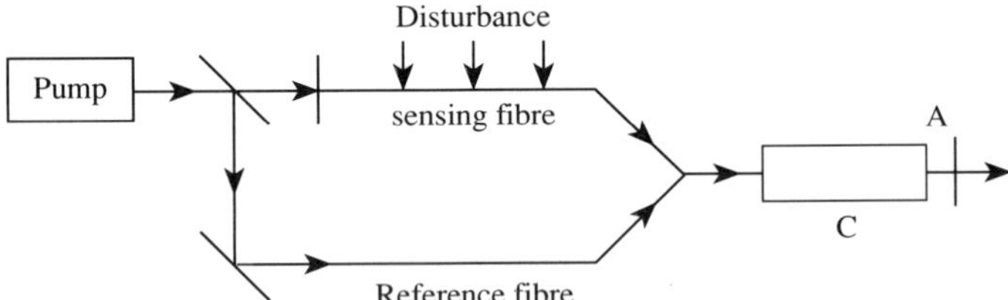

Fig. 1.27 The scheme of coherent measurement of small phase shifts $\varphi = (2\text{–}10) \times 10^{-2}$ rad between the waves in the sensing and reference fibres. The non-linear turning of the polarization ellipse formed in capillary C, and depending strongly upon the phase shift φ, results in a change of transmittivity of capillary C with analyser A at the output.

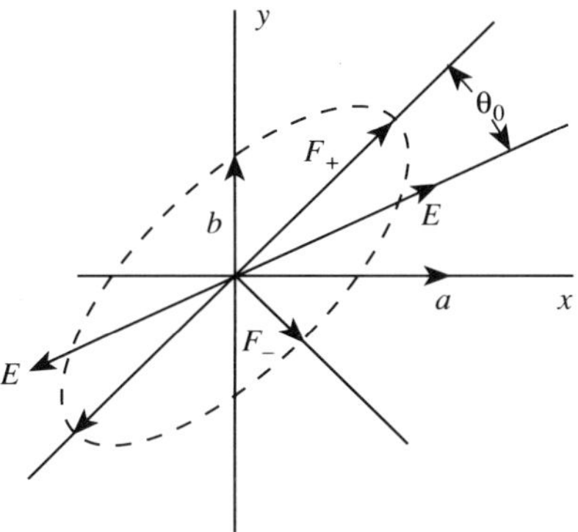

Fig. 1.28 The geometry of the ellipse formed at the input of capillary C (Fig. 1.27) due to phase shift φ between the orthogonal components of pumping radiation a and b. The phase shift being absent ($\varphi = 0$), the components a and b form a linearly polarized wave **E**. The non-zero phase shift results in formation of ellipse with half-axes F_+ and F_-. The orientation of the major half-axis F_+ is characterized by the angle θ_0 between F_+ and the direction of vector **E**.

orthogonal polarization components E_x and E_y is described by eqn (1.42). The directions of the components E_x and E_y, determined by polarizer P as well as their amplitudes $|E_x|_{z=0} = a$ and $|E_y|_{z=0} = b$, are fixed. The component E_y, passing through the sensing fibre, is characterized by a phase shift φ with respect to component E_x passing through the reference fibre. The straightforward use of the related theory of non-linear polarization advanced in Section 1.1 is impeded because the components E_x and E_y do not coincide with the principal axes of the polarization ellipse with the exception of the case $\theta_0 = \pm\pi/2$. Therefore, to apply this theory it is necessary to determine the principal axes of the polarization ellipse. Considering for definiteness that the condition $a > b$ is fulfilled, and using geometric transformations, one may write the major, $\mathbf{E}_1$, and minor, $\mathbf{E}_2$, axes of this ellipse in the form (Fig. 1.28)

$$E_{1,2} = aF_{1,2}; \qquad F_{1,2} = \sqrt{\tfrac{1}{2}(1 + \mu^2 \pm R)}; \mu = \frac{b}{a};$$

$$R = \sqrt{(1 + 2\mu^2 \cos 2\varphi + \mu^4)}. \tag{1.262}$$

The ratio of the axes of this ellipse is

$$\mu_0 = \left|\frac{E_2}{E_1}\right| = \left(\frac{1 + \mu^2 - R}{1 + \mu^2 + R}\right)^{1/2}. \tag{1.263}$$

The phase shift φ_0 between the field oscillations in the $\mathbf{E}_1$ and $\mathbf{E}_2$ directions is

$$\varphi_0 = \frac{\pi}{2} \operatorname{sign} \varphi; \qquad \operatorname{sign} \varphi = \begin{cases} +1, & \varphi > 0, \\ -1, & \varphi < 0. \end{cases} \qquad (1.264)$$

The angle θ_0 between the major axis E_1 and the fixed x_1-direction is related to the linear polarization of the unperturbed wave $(\varphi = 0)$ as

$$\theta_0 = -\tfrac{1}{2} \operatorname{sign} \varphi \, \arccos \left(\frac{1 + \mu^2 \cos^2 \varphi}{R} \right). \qquad (1.265)$$

The angle between the x and x_1-directions is $\delta_1 = \operatorname{arctg} \mu$. The wave's intensity $W_0 = a^2(1 + \mu^2)$ does not depend upon the phase φ. Since the parameter μ remains constant, the polarization state of the wave field at the input of the capillary, determined by parameters E_1, E_0, μ_0, J_0, and θ_0, depends only upon the phase shift φ, produced by perturbations of the sensing fibre. Thus the subsequent evolution of the polarization is determined by this phase φ. The angle of turning of the major axis $\mathbf{E}_1$ due to self-induced gyrotropy in the model (1.6) is

$$\theta = \frac{\pi}{2} \frac{z}{L_0} \operatorname{sign} \varphi; \qquad L_0 = \frac{\lambda_0}{2B\mu_0\chi a^2} \, . \qquad (1.266)$$

The parameter μ_0 is determined by (1.263). In the case of small values of phase shifts φ $(|\varphi| \ll 1 \, \mathrm{rad})$, the expressions for μ_0 (1.263), θ_0 (1.265) and L_0 (1.266) may be simplified;

$$\mu_0 = \frac{\mu|\varphi|}{1 + \mu^2} \, ; \qquad \theta_0 = -\frac{\mu^2\varphi}{1 + \mu^2} \operatorname{sign} \varphi; \qquad L_0 = \frac{\lambda_0(1 + \mu^2)}{4B\chi a^2|\varphi|} \, . \qquad (1.267)$$

Here the spatial scale L_0 proves to be in inverse proportion to the small parameter φ, and, as a consequence of eqn (1.266), we find θ to be proportional to φ. The growth of parameter φ leads to the weakening of such a dependence.

  This capillary together with the analyser A located at the output of the capillary $(z = L)$ acts as a non-linear polarization filter. The intensity of waves transmitted through this filter is characterize by the transmission coefficient K (1.261), which must be rewritten in conformity with our problem in the form

$$K = (1 + \mu^2)^{-1} \left[\cos^2(\theta + \theta_0 + \delta) + \mu_0^2 \sin^2(\theta + \theta_0 + \delta) \right]. \qquad (1.268)$$

Here θ is the angle of turning of the ellipse's major axis $\mathbf{E}_1$ from its initial direction $\theta = \theta_0$ (1.265), and δ is the angle between analyser's

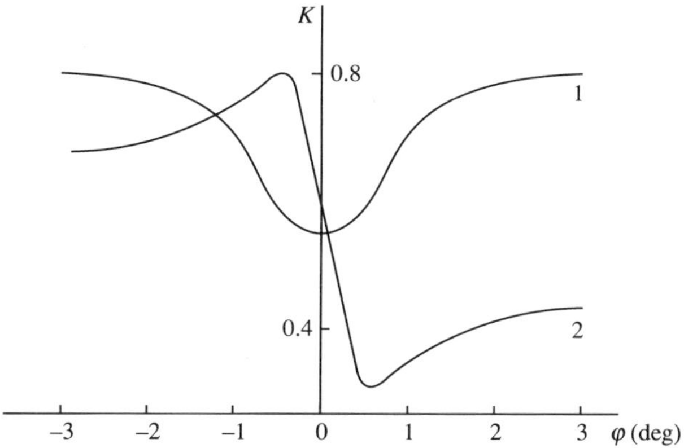

Fig. 1.29   Phase-dependent non-linear transmittivity K of a capillary waveguide (1.268) as a function of small phase shift φ between the components of the polarization ellipse at the waveguide's input. Curves 1 and 2 relate to the orientation of the transmission plane of the analyser A (1.27) characterized by the angles $\delta = 0$ and $\delta = -\pi/4$, respectively.

transmission plane and the x-direction at the waveguide's input.

Thus this device transforms the small phase shift at the input of the capillary into power modulations at its output. The examples of the characteristics of such a sensor presented in Fig. 1.29 exhibit high resolution in the range of small values of $|\varphi| = 1°\text{–}3°$. Observation, with the condition $\delta = 0$ fulfilled, permits us to measure only the value $|\varphi|$ (Fig. 1.29, curve 1, $K(\varphi) = K(-\varphi)$). Use of another condition, $\delta = -\pi/4$ (Fig. 1.29, curve 2, $K(\varphi) \neq K(-\varphi)$), illustrates the possibility of clarifying the sign of the phase perturbation and, moreover, of correcting simultaneously the value φ obtained in the previous case, $\delta = 0$. These graphs show the decrease of sensitivity connected with derivative $\partial K/\partial \varphi$ owing to growth of the phase shift φ.

Such a non-linear filter may be suitable for sensing of wide classes of perturbations, stimulating the appearance of small phase shifts φ. Thus, for example, this system may be used as a sensor for small temperature perturbations T_1 producing phase shifts in the sensing fibre (Fig. 1.27). The sensitivity of such a fibre is described by the ratio

$$p = \frac{1}{\beta l} \frac{|\varphi|}{T_1}. \qquad (1.269)$$

Here l is the length of sensing fibre and β is the wavenumber of the pumping wave in this fibre. Silica fibres without polymer coating may be characterized by a value of the parameter $p = 10^{-3}\,\mathrm{K}^{-1}$. Thus, measurement of a phase shift $|\varphi| = (1 \sim 2) \times 10^{-2}$ using a short span of fibre of $1 \sim 5\,\text{cm}$ permits us to detect temperature perturbations as small as $T_1 = 10^{-4}\,\text{K}$. The power of the pumping wave necessary for such measurement is $Q = W_0 S$, where S is the area of the capillary cross-section; in the case $\theta = \pi/4$, $L = 1.5\,\text{m}$, $S = 3 \times 10^{-6}\,\text{cm}^2$, one may evaluate the power as $Q \leqslant 30\,\text{kW}$. Using optical pulses of duration $T_0 = 1\,\mu\text{s}$ and choosing the repetition rate at about 10–20 pulses per second, connected with characteristic time of fibre cross-section heating of $\tau \leqslant 0.1\,\text{s}$, the energy expenditures for such sensing are of the order of magnitude 0.3–0.5 W.

Concluding this section we may remark that the high sensitivity in the range of small phase perturbations proves to be intrinsic also for other phase sensors based on the use of non-linear gyrotropic effects. The resolution of such systems may be improved according to eqn (1.257) in the case $\mu Q \sin \theta_0 = -1$; considering the limit $|\theta_0| = \pi/2 \pm \varphi$, $|\varphi| \ll \pi/2$, we find that parameter T tends to zero:

$$T = \frac{\varphi^2}{2}\,. \tag{1.270}$$

Thus a small variation of initial phase shift φ between interacting components of the polarization results in essential changes of polarization structure at the sensor's output. Unlike the linear dependence $z_0 \sim \varphi^{-1}$ ((1.267), $\theta_0 = 0$) for a non-gyrotropic medium, the quadratic dependence of gyrotropic sensors $z_0 \sim \varphi^{-2}$ (1.256, $\theta_0 = \pm\pi/2$) may provide better sensitivity of gyrotropic fibre sensors to small phase perturbations.

1.6 Tuning of powerful waves in single-core waveguide couplers

Interaction of modes described in the framework of the model of an ideal waveguide with the heterogeneities ignored is known to be limited by phase effects only. To analyse the evolution of mode amplitudes the waveguide's heterogeneities must be taken into account. Several examples of such amplitude–phase modulation of trapped waves provided by different heterogeneities of the guiding channels were discussed above:

1. spatial dispersion in single-core waveguide close to cutoff (Section 1.4);

2. tunnel penetration of modes between the channels of twin-core fibre (Section 1.5.1);

3. coupling of modes in a slightly gyrotropic single-core waveguide (Section 1.5.2).

The intensity-dependent phenomena in these coupling systems are produced by non-linear interaction of modes. The dynamics of this interaction depends upon the overlapping of localized modes. The spatial separation of guided modes in twin-core couplers restricts this overlapping only by interaction of evanescent tails of fields localized in each channel. In contrast, the single-core couplers are characterized by better overlapping of modes owing to close coincidence of their guiding channels. To design single-core couplers based on these improved overlapping properties one must understand the mechanism of mode coupling. It is noteworthy that the interaction of two modes in such couplers arises from artificial heterogeneities produced, for example, by the photorefractive effect, excitement of ultrasound waves in a waveguide, or corrugation of the core, while three-mode interaction may exist in single-core fibre without corrugation. The dynamics of two-mode phenomena is shown to depend the initial phase shift between modes; however, the interaction of modes proves to be independent of such phase shifts. These peculiarities determine the possibilities of using of such systems as phase-dependent sensors and power-dependent switches and filters.

1.6.1 *Phase-dependent non-linear amplifiers and sensors using corrugated waveguides*

Periodic variations of the dispersive properties of the waveguide's core result in dispersive coupling of propagating modes. These variations may be implemented either by modulation of the core's radius, the refractive index being unperturbed, or vice versa by modulation of the refractive index, the core's radius being unchanged. The first possibility is provided by special treatment of the core's surface, while the second method is connected with the photosensitivity of some fibre materials, which allows us to fix the phase grating within the core by holographic techniques. Although the coupling phenomena in both of these cases are based on fulfilment of the Bragg condition for propagation constants $\beta_{1,2}$ of interacting modes and the wavenumber q of the periodic structure,

$$\beta_1 - \beta_2 = q, \tag{1.271}$$

we may consider two limits related to different physical realizations of this equation:

$$|\beta_{1,2}| \gg |q|, \tag{1.272}$$

$$|\beta_{1,2}| - |q| \ll |\beta_{1,2}|. \tag{1.273}$$

This section is devoted to coupling of modes in the case (1.272); the opposite case (1.273) is discussed in Section 1.7.1.

It is worth stressing that the Bragg equation (1.271) may be combined with condition (1.273) only in the co-propagating geometry, and the formation of a reflected wave may be ignored in the further consideration of the corrugated waveguide. Such waveguide modulation, characterized by spatial period $\lambda_0 = 2\pi q^{-1}$, which is much longer than the wavelengths $\lambda_{1,2} = 2\pi\beta_{1,2}^{-1}$, may be accomplished by periodic variation of core's radius:

$$a = a_0 + a_1 \cos qz; \qquad a_1 \ll a_0. \qquad (1.274)$$

Considering for definitness the parabolically shaped waveguide profile (1.62) and using eqn (1.274), we obtain that the corrugation (1.274) results in variation of refractive index. The depth of modulation of the refractive index, n_1, is

$$n_1 = \frac{2n_0 a_1 \rho^2 \Delta}{a^3}. \qquad (1.275)$$

Making use of the resonant condition (1.271) we can write the coupling constant of co-propagating modes K produced by the waveguide's corrugation (1.274). To utilize the scheme of analysis of non-linear coupling of modes $\mathbf{E}_1$ and $\mathbf{E}_2$ advanced in Sections 1.4 and 1.5, this constant may be respresented as

$$p = \frac{\omega}{c}\, n_0 \langle n_1 (\mathbf{E}_1 \mathbf{E}_2) \rangle. \qquad (1.276)$$

Here the variation n_1 is given by (1.275); the symbol $\langle A \rangle$ indicates integration of function A over the waveguide's cross section:

$$\langle A \rangle = \frac{2\pi}{a^2} \int\limits_0^\infty A\rho\, \mathrm{d}\rho. \qquad (1.277)$$

Let us consider the definitness the coupling of Laguerre modes TE_{01} and TE_{02} in a parabolically shaped corrugated waveguide. Utilizing the standard notation

$$\mathbf{E}_1 = \mathbf{E}_{01} F_1 \mathrm{e}^{i\beta_1 z}; \qquad \mathbf{E}_2 = \mathbf{E}_{02} F_2 \mathrm{e}^{i\beta_2 z}, \qquad (1.278)$$

where the normalized distributions $\mathbf{E}_{01}$ and $\mathbf{E}_{02}$ relate to modes TE_{01} and TE_{02}, respectively, are given by (1.67). We obtain the system of equations governing the interaction of these fields in a form close to (1.164):

$$2i\beta \frac{\partial F_1}{\partial z} + \frac{\omega}{c}\, pF_2 \mathrm{e}^{i\gamma z} + \frac{\omega^2 \chi |M|^2}{c^2}\left[S_{11}|F_1|^2 + S_{12}|F_2|^2 \right] F_1 = 0,$$

$$(1.279)$$

$$2\mathrm{i}\beta\frac{\partial F_2}{\partial z} + \frac{\omega}{c}pF_1\mathrm{e}^{-\mathrm{i}\gamma z} + \frac{\omega^2\chi|M|^2}{c^2}\left[S_{21}|F_1|^2 + S_{22}|F_2|^2\right]F_2 = 0.$$

$$(1.280)$$

Here M is the amplitude of the TE_{01} mode at the waveguide's input ($z = 0$); the coefficients S_{mm} are connected with the overlapping of modes:

$$S_{mm} = \tfrac{3}{2}\left\langle|\mathbf{E}_{0m}|^4\right\rangle, \qquad (1.281)$$

$$S_{12} = S_{21} = \left\langle|\mathbf{E}_{01}|^2|\mathbf{E}_{02}|^2 + 2(\mathbf{E}_{01}\mathbf{E}_{02})^2\right\rangle. \qquad (1.282)$$

Substituting the distributions for the TE_{01} and TE_{02} modes (1.67) into eqns (1.281, 1.282) we find

$$S_{11} = S_{12} = S_{21} = \frac{3a^2}{8\pi\rho_0^2}; \qquad S_{22} = \frac{5}{8}S_{11}; \qquad p = \frac{4\omega\Delta}{\pi c}\frac{a_1}{a}n_0^2.$$

$$(1.283)$$

The dimensionless parameter ρ_0 is determined by (1.64), $\gamma = \beta_2 - \beta_1$.

Now the system (1.279, 1.280) may be analysed in close analogy with the problem of mode interaction in a twin-core coupler. Utilizing the dimensionless variable $\tau = pz\omega(2\beta c)^{-1}$ (1.166) and standard notation $F_{1,2} = f_{1,2}\exp(\mathrm{i}\varphi_{1,2})$, $f_{1,2} = |F_{1,2}|$, we can reduce the system (1.279, 1.280) to three equations governing the functions f_1, f_2 and φ with boundary conditions (1.171), while the equation describing the behaviour of the phase shift φ is distinguished from (1.170) by the difference between coefficients S_{22} and S_{11} (1.283):

$$\frac{\partial\varphi}{\partial\tau} = \frac{\cos\varphi(f_1^2 - f_2^2)}{f_1 f_2} - \frac{3R}{8}f_2^2, \qquad (1.284)$$

$$R = \frac{\omega\chi MS_{11}}{cp}. \qquad (1.285)$$

Since invariant (1.172) and eqn (1.174) are valid as before in the problem under discussion, we can derive the invariant

$$f_1 f_2\cos\varphi - \frac{3Rf_2^4}{32} = \mu\left(\cos\varphi_0 - \frac{3R\mu^3}{32}\right). \qquad (1.286)$$

Making use of eqns (1.169), (1.285) and (1.286) we obtain the equation determining the function f_2^2:

$$\frac{\partial^2 f_2^2}{\partial\tau^2} + 4f_2^2\left\{1 + \frac{3R\mu}{16}\left(\cos\varphi_0 - \frac{3R\mu^3}{32}\right)\right\} + \frac{9R^2}{128}f_2^6 = 2(1 + \mu^2).$$

$$(1.287)$$

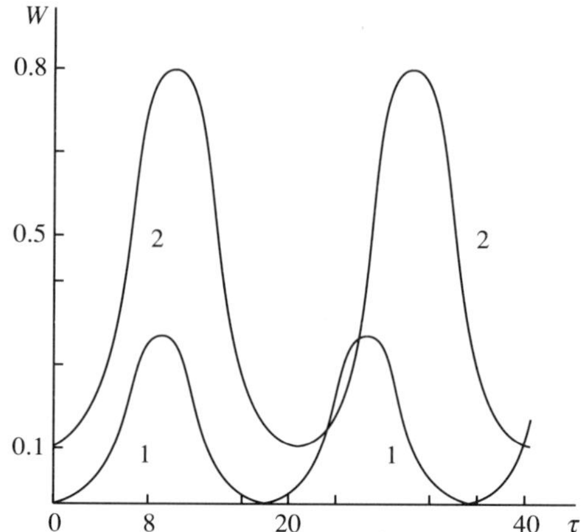

Fig. 1.30 Non-linear spatial oscillations of the dimensionless power of signal mode $W = I_1 I_0^{-1}$ due to transfer of power from the pumping mode I_0 in a corrugated waveguide. Curves 1 and 2 relate to the absence of signal ($W|_{\tau=0} = 0$) and to a small signal ($W|_{\tau=0} = 0.1$) at the waveguide's input $\tau = 0$, respectively.

The solution of this equation, similar to eqn (1.185), may also be presented by means of elliptical functions. This solution describes the power exchange between modes depending on the input power levels and initial phase shift between modes; thus the corrugated waveguide may be used as an effective intensity-dependent mode coupler.

To reveal the possibilities for such a corrugated fibre as a phase sensor it is expedient to reconsider the dynamics of mode coupling. Keeping in mind the typical sensor conditions and assuming the reference and sensor modes to be excited by powerful pumping and weak signal waves, respectively, let us examine the influence of a weak signal level μ and initial phase shift φ_0 on the maximum contrast of modes in the waveguide. Unlike in the previous analysis, we shall not restrict ourselves to the limiting cases $\varphi_0 = 0$ and $\varphi_0 = \pi$; instead, continuous variations of phase φ_0 will be analysed.

The dependence of spatial period of power oscillations in the signal mode upon the values of the parameter μ, given by eqn (1.287), is shown in Fig. 1.30. One can see that with the initial signal absent ($\mu = 0$, curve 1), the power in the signal mode tends to zero at the points $\tau = \tau_p$, determined by the intensity of pumping wave. The presence of a weak signal at the input of waveguide leads to formation of non-zero power flow at these points τ_p; this level is seen to exceed the initial level by 3–6 times. Thus the corrugated fibre may operate as an amplifier of weak

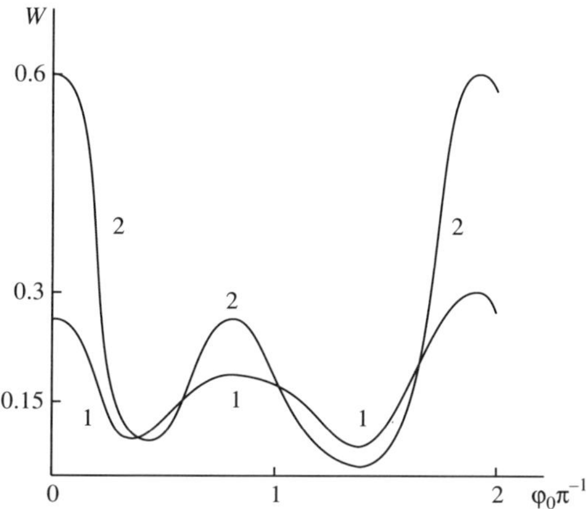

Fig. 1.31 The phase dependence of the amplification of a weak signal ($W|_{\tau=0} = 0.1$) intrinsic for energy transfer from the pumping mode in a corrugated fibre due to two-mode interaction. The difference of levels of power of the signal mode $\Delta W = W_{0.1} - W_0$ at the fibre's output $\tau = 2\tau_0$ corresponding to the cases $W_{0.1}|_{\tau=0} = 0.1$ and $W_0|_{\tau=0} = 0$ is plotted against the initial phase shift φ_0 between the signal and pumping modes at the fibre's input $\tau = 0$.

signals in all-optical systems, and this amplification appears attractive for recording of such signals in non-linear tunable devices.

The phase dependence of power level in the signal mode is illustrated in Fig. 1.31. The amplification of weak signals proves to be maximum in the range of small phase shifts $|\varphi| \ll 1$; the additional amplification of this signal, the phase shift being the same, may be achieved by doubling the length of the waveguide ($\tau = 2\tau_p$, curve 2). The phase sensitivity of amplification $K = K(\varphi_0)$ in the range $|\varphi_0| \leqslant \pi/2$ (Fig. 1.31) is about $|\partial K/\partial\varphi_0| = 0.5$ rad^{-1}, this value being approximately constant inside the wide range $|\varphi_0| \leqslant \pi/2$. Comparing this characteristic with the sensitivity of the phase sensor based of self-induced gyrotropy (Section 1.5.2, Fig. 1.29), one can see that the latter sensor may exhibit better sensitivity $|\partial K/\partial\varphi_0| = 4$–$5$ rad^{-1} in a narrow range of phase shifts $|\varphi_0| \leqslant 0.07$ rad^{-1}; out of this range the phase sensitivity of this sensor, $|\partial K/\partial\varphi_0| = 0.1$ rad^{-1} according to eqn (1.266), proves to be worse than the related characteristic of a corrugated waveguide.

Moreover, this phase-dependent amplification may be used for transformation of phase modulation into amplitude modulation in all-optical devices. The length of transformation determined by the period τ_p

(Fig. 1.30) in the fibre with $a = 2\,\mu\text{m}$, $a_1 a^{-1} = 10^{-15}$ $\delta = 0.04$, $\lambda_0 = 0.53\,\mu\text{m}$, pumping power $p = 1\,\text{W}$, is about $100\,\text{m}$.

It is necessary to stress that the phase sensitivity of mode coupling in corrugated waveguide, while useful for elaboration of sensors of amplifiers, may become pernicious in some other all-optical devices, such as power-dependent switches. Thus the important example of phase-independent coupling of modes is considered further.

1.6.2 *Phase-independent amplifiers and mode switches*

Unlike the above-mentioned effects of two-mode interactions in hetero-geneous or anisotropic waveguides, examples of three-mode coupling do exist even in homogeneous and isotropic single-core fibres. Such phenomena are produced by modal dispersion, when the propagation constants of interacting modes β_1, β_2, and β_3 obey the following condition:

$$\beta_1 + \beta_3 = 2\beta_2. \tag{1.288}$$

Let us discuss for definitness the cross-modulation of the TE_{01}, TE_{02}, and TE_{03} modes. The non-linear evolution of amplitudes of co-operating wave fields $F_{1,2,3} = f_{1,2,3} \exp\,(i\varphi_{1,2,3})$, $f_{1,2,3} = |F_{1,2,3}|$, the phase synchronism condition (1.288 being taken into account, is described by a system of equations similar to eqns (1.279, 1.280):

$$i\frac{\partial F_1}{\partial \tau} + F_1(u_{11}f_1^2 + u_{12}f_2^2 + u_{13}f_3^2) + u_0 F_3^* F_2^2 = 0; \tag{1.289}$$

$$i\frac{\partial F_2}{\partial \tau} + F_2(u_{21}f_1^2 + u_{22}f_2^2 + u_{23}f_3^2) + 2u_0 F_1 F_2^* F_3 = 0; \tag{1.290}$$

$$i\frac{\partial F_3}{\partial \tau} + F_3(u_{31}f_1^2 + u_{32}f_2^2 + u_{33}f_3^2) + u_0 F_1^* F_2^2 = 0. \tag{1.291}$$

The matrix elements u_{mn} are determined by the modes overlapping:

$$u_{mn} = \frac{S_{mn}}{S_{22}}\,; \qquad u_0 = \frac{S_0}{S_{22}}\,; \tag{1.292}$$

$$S_{mn} = \tfrac{3}{2}\langle|\mathbf{E}_{m\perp}|^4\rangle\,; \; S_{m=n} = \langle|\mathbf{E}_{m\perp}|^2|\mathbf{E}_{n\perp}|^2 + 2(\mathbf{E}_{m\perp}\mathbf{E}_{n\perp})^2\rangle\,; \tag{1.293}$$

$$S_0 = \langle(\mathbf{E}_{1\perp}\mathbf{E}_{2\perp})(\mathbf{E}_{2\perp}\mathbf{E}_{3\perp}) + \tfrac{1}{2}|\mathbf{E}_{2\perp}|^2(\mathbf{E}_{1\perp}\mathbf{E}_{31})\rangle\,. \tag{1.294}$$

Calculating these elements by means of distributions (1.67) related to the TE_{01}, TE_{02}, and TE_{03} modes we find

$$u_{11} = u_{12} = u_{21} = 1; \qquad u_{13} = u_{23} = u_{31} = u_{32} = \frac{3}{4};$$

$$u_{22} = \frac{5}{8}; \qquad u_{33} = \frac{15}{32}; \qquad u_0 = \frac{\sqrt{3}}{8}. \tag{1.295}$$

The variable τ is the normalized coordinate in the direction of wave propagation; since the linear coupling of modes is not discussed here, the non-linear length L_n is used for normalization:

$$\tau = zL_n^{-1}; \qquad L_n = \frac{16\pi\beta_2^2 c^2 \rho_0^2}{3\omega^2 a^2 \chi |M_2|^2}. \tag{1.296}$$

Here M_2 is the maximum amplitude of the TE_{02} mode at the fibre's input, and the scale ρ_0 is determined by (1.64).

Let us consider the boundary conditions related to the presence of a powerful pumping mode TE_{02} and a weak signal mode TE_{01} at the waveguide input, the TE_{03} mode being absent at the input and appearing in the course of evolution of the TE_{01} and TE_{02} modes:

$$f_1|_{\tau=0} - \mu; \qquad f_2|_{\tau=0} = 1; \qquad f_3|_{\tau=0} = 0. \tag{1.297}$$

The factor of importance for the system of eqns (1.289–1.291) is the coefficient u_0, connected with synchronous interaction of modes. Whereas the traditional non-linear terms that contain the sum of squares $f_{1,2,3}^2$ are known to determine the intensity-dependent phase shifts in this system, the new terms, which are proportional to u_0, provide the power transfer between modes. This process is characterized by two conservation laws characterizing the transfer of power in each pair of interacting modes:

$$2f_1^2 + f_2^2 = 1 + 2\mu^2, \tag{1.298}$$

$$2f_3^2 + f_2^2 = 1. \tag{1.299}$$

The addition of eqns (1.298) and (1.299) results in a trivial conservation law including all the interacting modes $W_{1,2,3} = f_{1,2,3}$:

$$W_1 + W_2 + W_3 = 1 + \mu^2. \tag{1.300}$$

Proceeding in a standard fashion we may write the non-linear equation describing the function W_2 in the limit $f_{30} = f_3|_{\tau=0}$, the values of parameters $u_{m,n}$ (1.295) being utilized:

$$\left(\frac{\partial W_2}{\partial \tau}\right)^2 = \frac{3}{16} W_2(1 + 2\mu^2 - W_2)(1 - W_2)$$

$$- \left[K_1 - \frac{(1 - W_2)(95 - 32\mu^2 - 129 W_2)}{128}\right]^2 \tag{1.301}$$

$$K_1 = \frac{\sqrt{3}}{2}\, \mu f_{30}\, \cos(2\varphi_{20} - \varphi_{10} - \varphi_{30}). \qquad (1.302)$$

All the coefficients in eqn (1.301) except K tend in the limit $f_{30} = 0$ to non-zero values, and the phase dependence of mode evolution is determined only by parameter K; hence it appears that the absence of the TE_{03} mode at the waveguide's input ($f_{30} = 0$, $K_1 = 0$) results in independence of three-mode interaction on the initial phase shifts between the TE_{01} and TE_{02} modes. Although the analytical solutions of eqn (1.301) may be expressed in terms of elliptical functions, it is convenient to illustrate some properties of these solutions using the corresponding graphs. Limiting ourselves to optical amplifier-related applications of these phenomena, we can observe in Fig. 1.32 the growth of an initially weak signal in the three-mode system. Increase of intensity of the weak signal is shown to be impeded due to simultaneous conversion of the pumping mode into the TE_{03} mode. This parasitic conversion accompanied by energy losses of the pumping wave stops the growth of the signal mode in some point τ_c and causes it to begin to decrease. The influence of the increasing TE_{03} mode at points $\tau > \tau_c$ becomes pernicious, and it is advantageous to switch out this parasitic mode using, for example, narrowing of the waveguide to stimulate the cutoff of the TE_{03} mode. The subsequent expansion of the waveguide's core up to the initial radius induces the formation of a new amplification cascade accompanied again by generation of a

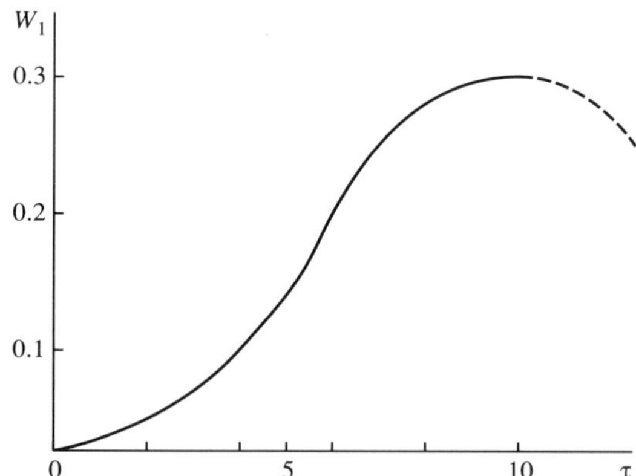

Fig. 1.32 The phase-independent growth of power W_1 of a weak TE_{01} signal simultaneously with growth of a parasitic TE_{03} mode due to energy transfer from a pumping TE_{02} mode in a corrugated waveguide ($W_1|_{\tau=0} = 0.1$). The dasted line shows the decrease of the level of the TE_{01} mode due to interaction with the parasitic TE_{03} mode.

parasitic TE_{03} mode, and so on. The amplification coefficient $W_1 W_{10}^{-1}$ in this twin cascade scheme shown on Fig. 1.31 exhibits considerable growth in the range of small input signals $W_1 W_{10}^{-1} \geqslant 5$.

The non-linear length L_n (1.296) evaluated for a parabolically shaped fibre with $a = 2\,\mu\text{m}$, $\delta = 0.04$, $\lambda_0 = 0.53\,\mu\text{m}$ is $L_n = 15\text{--}18\,\text{m}$, the launched power being about $1\,\text{W}$. It appears for Fig. 1.32 that the waveguide length necessary for one cascade amplification ($\tau_c = 10$) is $z = 150\text{--}180\,\text{m}$. The amplification process proves to be more effective in the range of small input signals. Thus this three-mode coupler may be used for intensity-dependent and phase-independent tuning of amplification and switching processes in waveguide all-optical circuits.

1.7 Tunable non-reciprocal dispersive elements in waveguide circuits

Non-linearly induced non-reciprocity of wave fields characterized in the simplest case by eqn (1.24) arises due to cross-interaction of two waves via the non-linear index grating that they generate. The manifestations of this effect have been studied extensively in Sections 1.5 and 1.6 devoted to coupling of co-propagating modes. Here the non-reciprocity of co-propagating waves in single-core waveguide is shown to exist due to difference of propagation constants caused by mode dispersion. However, the counter-propagating fields can exhibit non-reciprocal properties, the propagation constants being equal. These properties may become an effective tool for tuning of wide classes of wave fields with equal propagation constants, produced, for instance, by splitting of an initial wave beam. Thus, the optical systems based on counter-propagating geometry were proposed for the purpose of enhancement of the Sagnac effect and mutual self-bending of light beams (Kaplan 1981).

Although non-reciprocal phenomena do exist even in unbounded homogeneous non-linear media (Section 1.1) this section is devoted to peculiarities of these phenomena in dispersive and guided systems. Choosing the waveguide propagation regimes far from cutoff (Section 1.7.1) and close to cutoff (Section 1.7.2) we shall illustrate the influence of refractive index modulation and geometric dispersion on tuning of non-reciprocity of interacting wave fields.

1.7.1 *Non-linear coupling of counter-propagating modes in distributed Bragg filters*

Dispersional fibre filters are used for spatial separation of radiation flows with close wavelengths. Periodical variations of the waveguide's core were utilized first for elaboration of non-linear distributed Bragg filters (Stolen and Tom 1987). The basic components of these optical elements are phase gratings, which are imbedded within the cores of fibres. Germania (GeO_2)-doped silica fibres are known to possess a photosensitivity that permits use of the holographic technique for creation of such gratings. The refractive index of the fibre core due to this grating formation is characterized by spatial modulation

$$n = n_0 + n_1 \cos(2\beta z) \tag{1.303}$$

where $n_1 \ll n_0$, $\beta_0 = 2\pi n_0 \lambda_0^{-1}$, and λ_0 is the vacuum wavelength. The variation of refractive index (1.303) can provide an effective dispersional coupling of counter-propagating waves with equal frequencies ω and propagation constants $\pm\beta$. Since such a waveguide can be considered as a Bragg filter, its properties are highly dependent on the radiation wavelength. The transmission of this filter is determined by the phase distribution along the waveguide, this distribution being caused by both spatial dispersion of periodical structure and power-dependent self-phase modulation. In contrast to the coupling of co-propagating modes with slightly different propagation constants in a fibre with a corrugated core characterized by condition (1.272), the problem under discussion is distinguished by the possibility of coupling of counter-propagating modes due to resonant condition (1.273). This possibility is analysed below.

The dispersive coupling between forward $\mathbf{E}_1$ and backward $\mathbf{E}_2$ CW linearly polarized waves $(\mathbf{E}_1 \| \mathbf{E}_2)$ in a waveguide with spatially periodic refractive index (1.303) is described by the pair of equations (Winful 1985):

$$2i\beta_0 \frac{\partial E_1}{\partial z} + \frac{\omega}{c} n_1 \beta_0 E_2 e^{-2i\delta z} + \frac{\omega^2}{c^2} \chi(|E_1|^2 + 2|E_2|^2) E_1 = 0, \tag{1.304}$$

$$-2i\beta_0 \frac{\partial E_2}{\partial z} + \frac{\omega}{c} n_1 \beta_0 E_1 e^{2i\delta z} + \frac{\omega^2}{c^2} \chi(|E_2|^2 + 2|E_1|^2) E_2 = 0. \tag{1.305}$$

Here n_1 is the amplitude of spatial modulation of refractive index (1.303); the detuning parameter δ is connected with the wavenumber of this modulation:

$$\Delta = \frac{\omega n_0(\omega)}{c} - \beta \tag{1.306}$$

Unlike the equations governing the coupling of co-propagating waves (Sections 1.5 and 1.6), the negative sign in eqn (1.305) appears as a consequence of counter-propagating geometry. Introducing as usual the functions $E_{1,2} = af_{1,2} \exp(i\varphi_{1,2})$ and

$$\theta = \varphi_2 - \varphi_1 - 2\Delta z \tag{1.307}$$

we may obtain from (1.304, 1.305) the dimensionless system of equations:

$$\frac{\partial f_1}{\partial \tau} = -f_2 \sin \theta; \tag{1.308}$$

$$\frac{\partial f_2}{\partial \tau} = -f_1 \sin \theta; \tag{1.309}$$

$$\frac{\partial \theta}{\partial \tau} = -2L_0\Delta - (f_1^2 + f_2^2)\left[\frac{\cos \theta}{f_1 f_2} + 3\gamma\right]. \tag{1.310}$$

Here the normalized distance τ is connected with scaling length L_0:

$$\tau = zL_0^{-1}; \qquad L_0 = \frac{2c}{\omega n_1}. \tag{1.311}$$

Parameter γ is determined by non-linearity and the waveguide's spatial modulation:

$$\gamma = \frac{\chi a^2}{n_0 n_1} \tag{1.312}$$

The system of equations (1.308–1.310) describes the power-dependent reflection and transmission of the wave interacting with a Bragg distributed filter. Supposing the intensity of the incident wave at the filter's input $\tau = 0$ to be equal to 1, and designating the intensity reflection coefficient as μ^2,

$$f_1^2\big|_{\tau=0} = 1; \qquad f_2^2\big|_{\tau=0} = \mu^2, \tag{1.313}$$

we can conclude from eqns (1.308, 1.309) the conservation law

$$f_1^2 - f_2^2 = 1 - \mu^2, \tag{1.314}$$

which differs from the related law for co-propagating modes: $f_1^2 + f_2^2 = 1 + \mu^2$. Here the intensities of these waves at the filter's output $\tau = \tau_0$ are

$$f_1^2\big|_{\tau=\tau_0} = 1 - \mu^2; \qquad f_2^2\big|_{\tau=\tau_0} = 0. \tag{1.315}$$

Another conservation law of the system (1.308–1.310) may be written as

$$f_1 f_2 \cos \theta + \frac{3\gamma}{2}(f_1 f_2)^2 + \frac{L_0\Delta}{2}(f_1^2 + f_2^2) = \frac{L_0\Delta(1 - \mu^2)}{2}. \tag{1.316}$$

To solve the set of eqn (1.308–1.310) let us derive two auxiliary equations,

$$\frac{\partial (f_1^2 + f_2^2)}{\partial \tau} = -4f_1 f_2 \sin \theta, \tag{1.317}$$

$$\frac{\partial (f_1 f_2)}{\partial \tau} = -(f_1^2 + f_2^2) \sin \theta. \tag{1.318}$$

Introducing the new function

$$V = f_1^2 - \frac{1 - \mu^2}{2} \tag{1.319}$$

and making use of the identity

$$(f_1 f_2)^2 = V^2 - \frac{(1 - \mu^2)^2}{4} \tag{1.320}$$

we may find the equation governing the function V:

$$\frac{1}{4} \frac{\partial^2 V}{\partial \tau^2} - V \left\{ 1 - (L_0 \Delta)^2 + \frac{3\gamma (1 - \mu^2)}{2} \left[L_0 \Delta + \frac{3\gamma}{4} (1 - \mu^2) \right] \right\}$$
$$+ \frac{9\gamma (L_0 \Delta)^2}{2} V^2 + \frac{9\gamma^2}{2} V^3 - \frac{(L_0 \Delta)(1 - \mu^2)}{2} \left[L_0 \Delta + \frac{3\gamma (1 - \mu^2)}{4} \right] = 0. \tag{1.321}$$

To obtain the first integral of eqn (1.321) we may take into account the following properties of function V:

$$V\big|_{\tau = \tau_0} = \frac{1 - \mu^2}{2}; \qquad \frac{\partial V}{\partial \tau}\bigg|_{\tau = \tau_0} = 0 \tag{1.322}$$

and introduce the function t by means of the relation

$$V = \frac{(1 - \mu^2)}{2} \left(1 + \frac{1}{t} \right). \tag{1.323}$$

Since $V\big|_{\tau = 0} = (1 + \mu^2)/2$, the function t is increasing from the value $t\big|_{\tau = 0} = t_0$,

$$t_0 = \frac{1 - \mu^2}{2\mu^2}, \tag{1.324}$$

up to the limit $t\big|_{\tau = \tau_0} \to \infty$. The use of the function t permits us to write the first integral of eqn (1.321) in a simple form convenient for the subsequent analysis:

$$2\tau = -\int_{\infty}^{t} \frac{dt}{\sqrt{\{2t^3 + t^2 - [(b + 2q)t + q]^2\}^{1/2}}}; \tag{1.325}$$

$$b = L_0 \Delta; \qquad q = \frac{3\gamma(1 - \mu^2)}{4}. \qquad (1.326)$$

Since the denominator in eqn (1.325) contains a cubic polynomial, this integral can be expressed in terms of elliptic integrals of the first kind. The roots of this polynomial depend on two factors: the factor b describing the spectral width of filtering waves, and the factor q connected with non-linear and dispersive properties of the filter. It should be emphasized, that the characteristics of both these last two

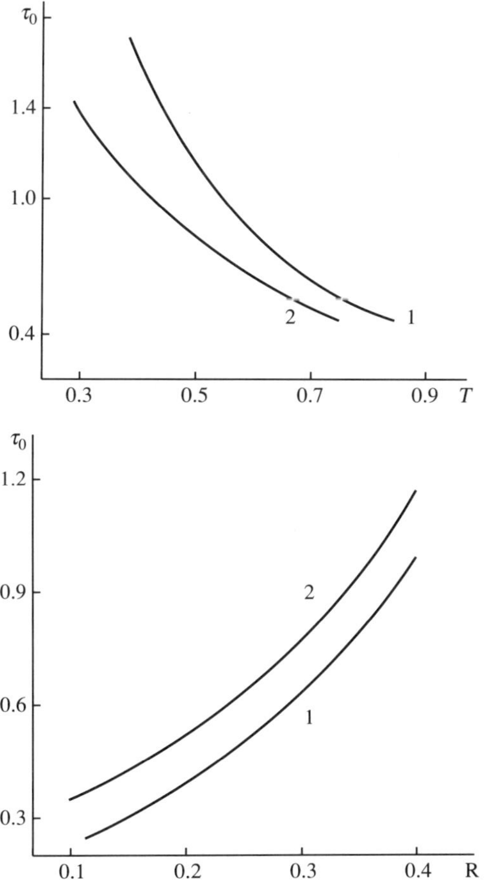

Fig. 1.33 Non-linear tuning of distributed Bragg filters (1.303) for counter-propagating waves. The lengths of filters τ_0 are given as functions of the transmission T (a) and reflection R (b) coefficients. Curves 1 and 2 relate to the values of the non-linear and dispersive parameters $q = 0.06$, $b = +1$ (1.326) and $q = 0.32$, $b = -1$ (a), and $q = 0.77$, $b = -1$ and $q = 0.17$, $b = +1$ (b) respectively.

properties are combined in one parameter q (1.326).

Let us begin the analysis of eqn (1.325) from the low-intensity limit $\gamma = 0$ ($q = 0$). Thus, for example, in the case of ($b^2 < 1$) we obtain straightforwardly from (1.325) the well-known result of coupled-mode theory (Marcuse 1974) describing the intensity transmission coefficient I of such a filter:

$$T = 1 - \mu^2 = [1 + (1 - b^2)^{-1}\,\text{sh}^2(\tau_0\sqrt{1 - b^2})]^{-1} \qquad (1.327)$$

The case $b^2 > 1$ may be analysed analogously.

Consideration of the non-linear problem ($q \neq 0$) demands the calculation of roots of denominator in eqn (1.325). In a special case related to zero detuning ($b = 0$) these roots can be written readily as

$$t_{1,2} = q(q \pm \sqrt{1 + q^2}); \qquad t_3 = -\tfrac{1}{2}. \qquad (1.328)$$

Numerical calculation is needed for roots $t_{1,2,3}$ in the case of non-zero detuning. The quantities $t_{1,2,3}$ being known, the standard methods of calculation of elliptical integral may be utilized (Janke *et al.* 1960).

The power-dependent intensity transmission (T) and intensity reflection (R) coefficients for different lengths of fibre span (τ_0) are presented in Fig. 1.33. To obtain the given values of coefficients T and R, for fixed length τ_0, the related non-linear parameter γ must be determined from eqn (1.311). An example of intensity-dependent tunable transmittivity of the Bragg filter is illustrated in Fig. 1.34. The tuning is controlled by additional phase shift provided by non-linearity. Using

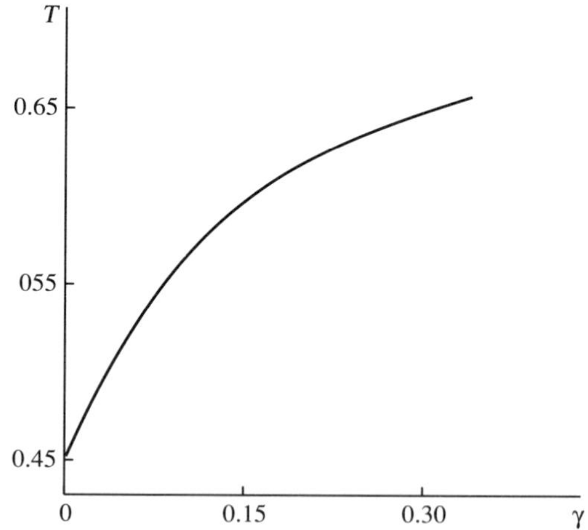

Fig. 1.34 Intensity-dependent tunable transmittivity of a Bragg filter (1.303) for counter-propagating waves ($\tau_0 = 0.9, b = +1$). The coefficient of transmittivity T is plotted against the non-linear parameter γ (1.312).

the typical value of modulation parameter $n_1 = 10^{-5}$ (1.303) one may conclude from eqn (1.311) that the typical length of such a Bragg filter for vacuum wavelengths $\lambda_0 = 1.3$–$1.5\,\mu\mathrm{m}$ is about $5 \sim 10\,\mathrm{cm}$. Thus non-linear Bragg filters seem to be applicable for tunable spectral filtration of radiation flows in waveguide circuits.

1.7.2 *Enhancement of non-reciprocal phase effects in waveguides close to cutoff*

The enhancement of phase self-modulation of powerful waves travelling in the waveguide close to the cutoff regime, discussed in Section 1.4, stimulates interest in non-reciprocal phenomena for such waves. Here the interaction of counter-propagating modes in such a regime leads to additional possibilities of tuning these modes. It is instructive to illustrate these possibilities by considering the interaction of fields which possess two or three polarization components. Using, for example, the TM_{01} modes (1.138) in the counterpropagating geometry in a model of a slab waveguide with metallic walls, we can present the polarization structure of interacting waves $\mathbf{E}_1$ and $\mathbf{E}_2$ in the form

$$E_{1x} = a_1 e^{i\beta z + i\varphi_1} \cos k_\perp x; \qquad E_{1z} = i k_\perp \beta^{-1} a_1 e^{i\beta_1 z + i\varphi_1} \sin k_\perp x,$$

$$(1.329)$$

$$E_{2x} = a_2 e^{-i\beta z + i\varphi_2} \cos k_\perp x; \qquad E_{1z} = -i k_\perp \beta^{-1} a_2 e^{-i\beta z + i\varphi_2} \sin k_\perp x.$$

$$(1.330)$$

Substituting these components into eqn (1.137) and proceeding by analogy with calculations (1.140–1.147), we find that the amplitudes of interacting waves a_1 and a_3 remain constant. The phases of interacting waves φ_1 and φ_2 vary, these variations being strongly dependent upon the parameter y^2 (1.141) connected with the proximity of the propagation regime to cutoff:

$$\varphi_1 = [(3 + 2y^2 + 3y^4)a_1^2 + 6(1 + y^4)a_2^2]\tau, \qquad (1.331)$$

$$\varphi_2 = [6(1 + y^4)a_1^2 + (3 + 2y^2 + 3y^4)a_2^2]\tau. \qquad (1.332)$$

The dimensionless distance τ is determined by (1.144).

The inequality of expressions (1.331) and (1.332) illustrates the non-reciprocity of phase perturbations of interacting TM_{01} modes. The relative phase shift

$$\varphi = \varphi_2 - \varphi_1 = (3 - 2y^2 + 3y^4)(a_1^2 - a_2^2)\tau \qquad (1.333)$$

diminishes slightly in the range $0 < y^2 \to 3^{-1}$ and is increasing when $y^2 > 3^{-1}$. Thus the utilization of propagation regimes close to cutoff

($y^2 > 1$) shows the possibility of amplifying by $(1 - \frac{2}{3} y^2 + y^4)$ times, in comparison with the case $y = 0$, the non-reciprocal phase shifts in devices operating with small phase shifts such as the Sagnac rotation sensor.

An example discussed above relates to interaction of two TM_{01} modes travelling in opposite directions. Here the relative phase shift φ (1.333), dependent upon the difference of mode intensity, vanishes, the intensities being equal. In contrast, the non-reciprocal phase effect produced by different counter-propagating modes does exist even for equal intensities of interacting modes. Considering, for example, TE_{01} and TE_{02} modes, counter-propagating in a parabolically shaped waveguide (1.62), the terms depending on the parameter $y^2 = k_\perp^2 \beta^{-2}$ being rigorously absent, we obtain

$$\varphi = \left(1 - \frac{3\mu^2}{8}\right)\tau_n. \tag{1.334}$$

Here $\mu = a_2 a_1^{-1}$ $a_{1,2}$, are the amplitudes of the TE_{01} and TE_{02} modes respectively; τ_n is the normalized distance $\tau_n = zL_n^{-1}$. L_n is the characteristic length of non-linear phase self-modulation for the TE_{01} mode,

$$L_n = \frac{\lambda_0}{4\sqrt{2\Delta}\,\chi|E_{01}|^2}. \tag{1.335}$$

This phase effect tends to zero when the amplitudes of interacting TE_{01} and TE_{02} modes obey the condition $a_1^2 = \frac{3}{8} a_2^2$.

1.8 Hysteretic variations of mode parameters in narrow-banded resonant systems

Unlike the continuous intensity-dependent variation of parameters of waveguide modes discussed above, such as polarization, phase shifts, and power distribution between modes, this section is devoted to possibilities of jump-like changes of these parameters. These jumps are important for tuning of radiation flows in all-optical switches and information systems, including logical gates and memory devices. The related set-up may be based on bistable phenomena occurring owing to power-dependent transmission and reflection of light. Postponing the description of the corresponding reflection phenomena to Chapter 2, we shall analyse here discontinuous changes of power, polarization, phase shifts, and mode coupling using non-linear transmission effects. Since jumps of the parameters of wave fields in guided system are

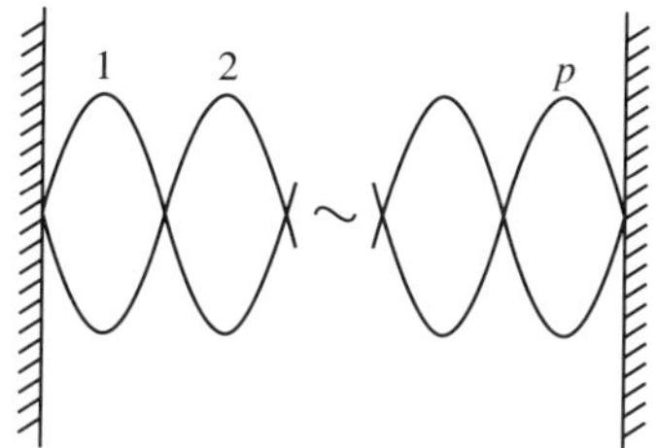

Fig. 1.35 A scheme of the optical Fabry–Perot cavity. The two mirrors are separated by p half-wavelengths.

produced by hysteresis of transmitted power, it is worth analysing first the conditions of appearance of this hysteresis.

The appearance of two power-dependent stable states of the transmitted field is usually demonstrated by means of a Fabry–Perot resonator filled with non-linear medium, e.g. sodium (Gibbs *et al.* 1976; Smith *et al.* 1978). The simplest resonator of this type is the cavity in which two mirrors oppose each other across the non-linear medium (Fig. 1.35). The eigenfrequencies of electromagnetic field oscillations in this cavity are

$$\Omega_p = \frac{\pi c p}{Ln}, \tag{1.336}$$

where p is an integer ($p = 1, 2, 3, \ldots$) and n is the average refractive index over the length of the cavity, L. This expression corresponds to p half-wavelengths fitting between the mirrors; each value of p relates to a different mode of the cavity. The transmittivity of the cavity is known to be dependent upon the mode reflection coefficient R and the round-trip phase φ. Neglecting for simplicity the absorption of the wave in the cavity, one may describe the transmittivity of the cavity using the formula (Gibbs 1985)

$$T = \left[1 + \frac{4R}{(1-R)^2} \sin^2 \frac{\varphi}{2} \right]^{-1}, \tag{1.337}$$

$$\varphi = \frac{4\pi Ln}{\lambda}. \tag{1.338}$$

Thus the transmittivity may vary between the limits

$$1 \leqslant T \leqslant \left(\frac{1-R}{1+R} \right)^2 \tag{1.339}$$

owing to alterations of the phase φ. The output intensity, I_{out}, and the intensity accumulated inside the resonator, I_R, are connected with the input intensity I_{in}:

$$I_{out} = TI_{in}; \qquad I_R = \frac{(1 - R)}{1 + R} TI_{in}. \qquad (1.340)$$

The intensity-dependent phase shift φ,

$$\varphi = \varphi_0 + \varphi_1, \qquad \varphi_0 = \frac{4\pi L n_0}{\lambda}, \qquad \varphi_1 = \frac{4\pi L n_2 I}{\lambda}, \qquad (1.341)$$

results in non-linear transmittivity $T = T(I_{in})$.

Although this approach forms the physical foundation of the series of optical devices, based on hysteretic properties of localized standing waves, we shall analyse here only some examples of related devices, such as Fabry–Perot and ring cavities. These bistable resonators, widely used in switching, sensing, and interferometric systems, exhibit the physically meaningful models of single-wave and double-wave pumping. Being connected with the eigenfequencies of resonant systems ($\omega = \Omega_p$ (1.336)), both these models prove to be spectrally narrow-banded.

1.8.1 *Jumps in transmittivity of optical circuits using Fabry–Perot waveguide resonators*

The appearance of multi-valued dependence between the input (P_{in}) and the output (P_{out}) powers in an optical system may be provided by the tunable hysteretic transmittance of a Fabry–Perot resonant cavity formed by the span of the waveguide. This hysteresis is connected with non-linear phase shift φ (1.341). To examine the region of multi-valued dependence of P_{out} on P_{in}, let us start with eqn (1.340). Making use of eqns (1.341) we obtain

$$\frac{\partial P_{out}}{\partial P_{in}} = \frac{T}{1 - \dfrac{\Delta\varphi}{T}\dfrac{\partial T}{\partial\varphi}} \qquad (1.342)$$

It is useful to describe the non-linear behaviour of a Fabry–Perot resonator by means of a curve of states as a function of input power I_{in} in the (I_{in}, I_{out}) plane and search the region in which this curve becomes multi-valued. The curve $I_{out} = I_{out}(I_{in})$ in this region proves to be Z-shaped, with three branches (Fig. 1.36). Such dependence exhibits bistability properties, wherein two branches relate to stable states while the third branch is considered to be unstable. Let us consider these two stable branches which have a continuation into the

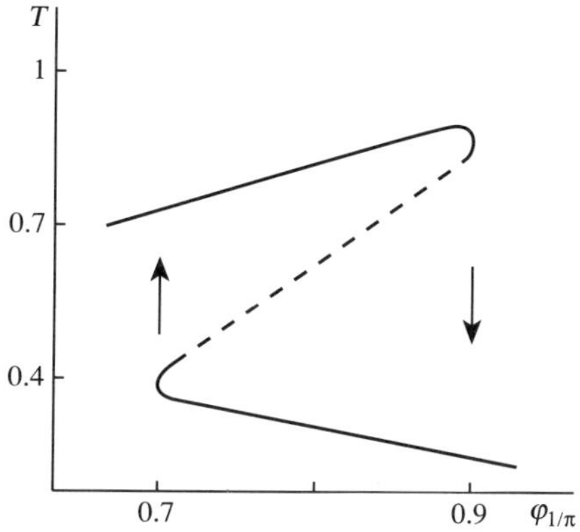

Fig. 1.36 Bistability of the transmission of a Fabry–Perot cavity $T = T(\varphi_1)$, where φ_1 is given by eqn (1.341). The critical points of the graph $T = T(\varphi_1)$ relate to Fig. (1.37).

region of a single state (the solid branches of curve P_{out} in Fig. 1.36). The transition jumps between stable states should take place at points P_{in1} and P_{in2} when the derivative $\partial P_{\text{out}}/\partial P_{\text{in}}$ tends to infinity; the signs of derivative $\partial P_{\text{out}}/\partial P_{\text{in}}$ inside and outside $(P_{\text{in1}}, P_{\text{in2}})$ segments are opposite. These points relate to zero value of the denominator in eqn (1.342):

$$\frac{\partial T}{\partial \varphi} = \frac{T}{\Delta \varphi} \, . \tag{1.343}$$

It is worthy analysing some solutions of eqn (1.343) graphically (Ferber and Marburger 1976) by means of Fig. 1.37. Since the condition (1.343) determines the tangents of the curve $\tau = \tau(\Delta\varphi)$ passing at the $(\tau, \Delta\varphi)$ plane through the point $(0,0)$, one may find several pairs of solutions of eqn (1.343) characterizing the transmittivities $\tau_{1,2}$ and related phase shifts $\Delta\varphi_{1,2}$ of both stable states (Fig. 1.37). Each of these pairs of solutions describes the corresponding Z-shaped part of the multi-stable curve of states. The parameters of critical points $P_{\text{in1,2}}$ and $P_{\text{out1,2}} = T_{1,2}P_{\text{in1,2}}$ may be calculated by substitution of related values $\varphi_{1,2}$ and $T_{1,2}$ into eqn (1.337).

An example of several hysteretic loops typical of non-linear behaviour of Fabry–Perot waveguide resonators is illustrated in Fig. 1.38. These energy-dependent transitions between the stable states indicated by arrows are shown to relate to jump-like changes of transmitted power P_{out} by a factor of 2–3. Thus this resonator may be used both for

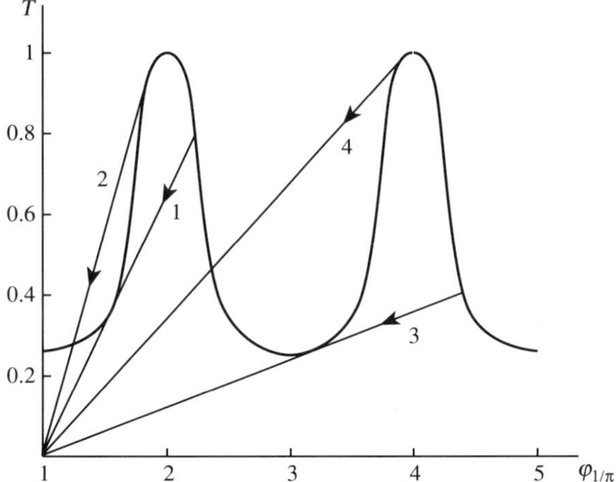

Fig. 1.37 The curve T is a plot of Fabry–Perot transmission as a function of non-linear phase shift φ. The straight lines 1, 2, 3, and 4 are tangents to the curve T satisfying eqn (1.343). The lines 1 and 3 relate to jumps up between the stable branches of the S-like graph $P_{\text{out}} = P_{\text{out}}(P_{\text{in}})$ (Fig. 1.36), while the lines 2 and 4 correspond to jumps down. The lines 1, 2 and 3, 4 describe the first and second bistability loops, respectively. The parameters of the waveguide resonator formed by central strip ($n_0 = 3.6$, $n_2 = 6 \times 10^{-14}\,\text{V}^2\,\text{m}^{-2}$) surrounded by layers with $n_1 = 3.4$ are $L = 1.3 \times 10^{-2}\,\text{cm}$, $2a = 0.225\,\mu\text{m}$, $R = 0.3$.

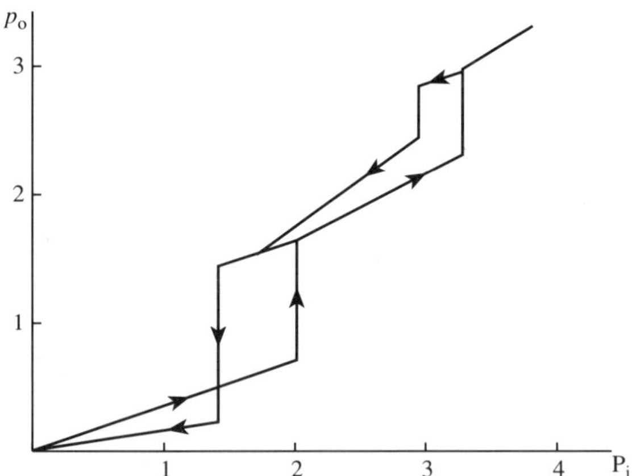

Fig. 1.38 Optical multi-stability of the TE_{01} mode in a slab waveguide resonator; the resonator's parameters are given at the caption to Fig. 1.37;
$p_{\text{i}} = \ln\left(\dfrac{P_{\text{in}}}{10^3}\right)$; $p_{\text{o}} = \ln\left(\dfrac{P_{\text{out}}}{10^3}\right)$; both P_{in} and P_{out} are given in $W\,\text{cm}^{-1}$.

high-speed switching of power flows and for discrimination of small powers. This jump-like transmission can provide the tuning of resonator's signals, which may be used further as input signals in the above-mentioned non-linear fibre systems for polarization, modulation, and coupling of modes (Sections 1.5–1.7). Being excited by hysteretic states of the pumping beam, characterized by intensity parameter, these systems are also acting in bistable regimes.

It is while pointing out that these phenomena are caused by self-action of single pumping beam. In contrast, the hysteretic non-reciprocal effect resulting from cross-interaction of two pumping beams inside the resonant cavity is analysed below.

1.8.2 *Field bistability in a ring resonator*

The cross-interaction of two counter-propagating waves was shown to give rise to non-reciprocal perturbations of the waves' phases (1.24) via the non-linear index grating, the wave amplitudes being unperturbed. A similar effect produces the bistable behaviour of amplitudes of such waves inside a non-linear ring resonator symmetrically pumped by two beams of the same frequency and intensity (Fig. 1.39). This effect could not occur in the above-mentioned case of single-wave pumping.

To find the conditions for this bistability to arise let us take into account the shift of the resonator's eigenfrequency Ω_p (1.336) due to insertion of a thin non-linear plate–into the resonator as shown in Fig. 1.39. Denoting the total optical length of the resonator as L, and the optical length of non-linear plate as L_n, ($L_n \ll L$), one can obtain from (1.336)

$$\Omega_p = \Omega_{p0} - \Omega_{p0}\frac{L_n}{L}\Delta\mathscr{L}_N. \tag{1.344}$$

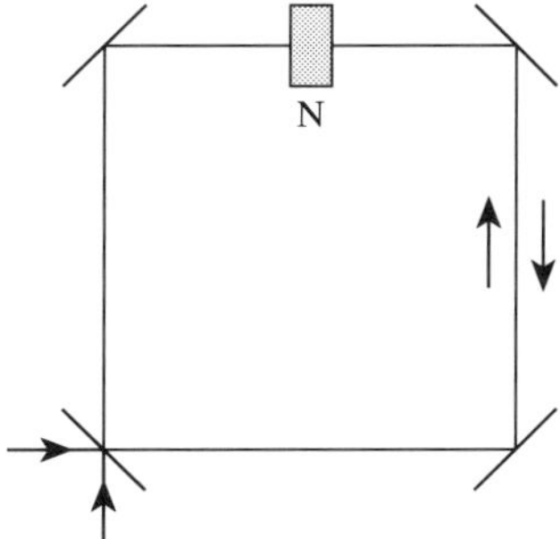

Fig. 1.39 A ring resonator symmetrically pumped by two laser beams. N is the non-linear element inserted into the resonator.

Here Ω_{p0} is the unperturbed eigenfrequency, $\Delta\mathscr{E}_N$ is the non-linear component of susceptibility of the plate M. This component $\Delta\mathscr{E}_{Nj}$ $(j = 1, 2)$ was shown to be non-reciprocal for a Kerr-like medium (1.124). Taking the frequency ω of the pumping beams to be close to one of the resonator's eigenfrequencies Ω_p, we can express the steady-state amplitude E_j $(j = 1, 2)$ inside the resonator via the incident field E_{in} (Kaplan 1981):

$$E_j = \frac{-iE_{in}\gamma_c T^{-1}}{\omega - \Omega_p + \Omega_p(L_n/L)\Delta\mathscr{E}_{Nj} - i\gamma} \ . \tag{1.345}$$

Here T is the mirror transmittivity; $\gamma_c = cTL^{-1}$ is the empty cavity bandwidth; parameter $\gamma = \gamma_c + \gamma_s$ gives the total bandwidth of the system; γ_s gives the linear losses of the medium inside the cavity. Let us introduce the dimensionless detuning

$$\Delta = \frac{\omega - \Omega_p}{\gamma} \tag{1.346}$$

and normalized intensities

$$I_j = \frac{\mathscr{E}_2|E_j|^2 L_n\Omega_p}{L\gamma} \ ; \qquad A = \frac{\mathscr{E}_2|E_{in}|^2 L_n\Omega_p\gamma_c^2}{L\gamma^3 T} \ . \tag{1.347}$$

Multiplying eqn (1.345) by its complex conjugate and assuming the intensities of both fields I_1 and I_2 excited by the same pumping beams to be equal $(I_1 = I_2 = I_0)$, we obtain the dimensionless equation governing the field I_0 in the ring resonator:

$$A = I_0[1 + (\Delta + 3I_0)^2] . \tag{1.348}$$

The dependence of intensity I_0 upon the detuning Δ described by eqn. (1.348) for a given the pumping intensity A exhibits hysteretic behaviour. The graphic method (Fig. 1.37), convenient for harmonic functions, is not valid here, and we shall use the standard analytical procedure. By requiring that

$$\frac{\partial\Delta}{\partial I_0} = 0 \tag{1.349}$$

at the points of hysteretic jumps, we find readily from eqn. (1.348) that the bistability appears at the points $I_{01,2}$:

$$I_{01,2} = \tfrac{1}{9}\left(-2\Delta \pm \sqrt{\Delta^2 - 3}\right). \tag{1.350}$$

The detuning threshold for the onset of bistability is $\Delta_t = -\sqrt{3}$, and the condition of bistability appearance is connected with a sufficient degree of detuning:

$$\Delta < \Delta_t = -\sqrt{3}. \tag{1.351}$$

Substituting the value $\Delta = \Delta_t$ into eqns (1.350) and (1.348), we obtain the related threshold intensity $A_t = 8/(9\sqrt{3})$. The intensity of the pumping beam characterized by the values of parameter $A < A_t$ corresponds to monotonic dependence $I = I(\Delta)$, whereas more powerful beams $(A > A_t)$ provide the formation of two stable states of field I_0 inside the resonator (Fig. 1.40).

This analysis is valid as long as the polarization of counter-propagating waves E_j is supposed to be linear. The more sophisticated situation is connected with the interaction of two elliptically polarized waves in the same ring resonator (Fig. 1.39). This effect is shown to depend upon the phase shifts θ_j between the polarization components of each wave. In a particular case, when the polarization of bothwaves is linear $(\theta_1 = \theta_2 = 0)$, the interaction discussed results in formation of power-dependent index grating, which causes the non-reciprokal phase shifts, the wave amplitudes remain invariant (see Section 1.1). In contrast, the cross modulation of elliptically polarized waves $(\theta_1 \neq 0;$ $\theta_2 \neq 0)$ provides the perturbations of their polarization components, the total power of each wave being conserved. This phenomenon proves to be useful for measurements of small phase shifts θ_j via the observations of perturbed polarization states of both fields. The sensitivity of phase sensor based on this method may be enhanced noticeably due

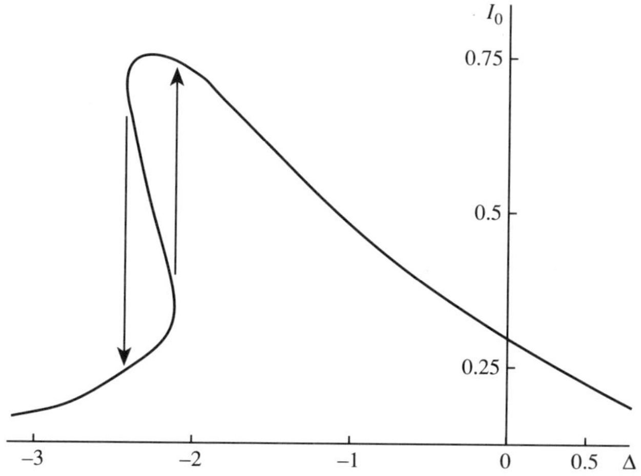

Fig. 1.40 Bistable dependence of dimensionless power I_0 inside the non-linear resonator upon the frequency detuning Δ; the dimensionless pumping power $A = 0.5$ (Eq. 1.347).

to using of non-linear resonant properties of the circuit shown on Fig. 1.39.

To emphasize this possibility let us discuss the pair of elliptically polarized counter-propagating waves $\vec{E}_1$ and $\vec{E}_2$

$$\vec{E}_1 = e^{ikz}(\vec{n}_1\alpha_1 e^{i\theta_1/2} + \vec{n}_2 b_1 e^{-i\theta_1/2})$$
$$\vec{E}_2 = e^{-ikz}(\vec{n}_1\alpha_2 e^{-i\theta_2/2} + \vec{n}_2 b_2 e^{i\theta_2/2})$$

$$(1.352)$$

Here $\vec{n}_1$ and $\vec{n}_2$ are the orthogonal polarization vectors directed along the axes X and Y respectively. Since the fields $\vec{E}_1$ and $\vec{E}_2$ are formed due to splitting of an initial elliptically polarized wave beam (Fig. 1.39) one can suppose for simplicity, that $\theta_1 = \theta_2 = \theta$. Substituing the expressions (1.352) into Eqn (1.26) we find the non-linear electric displacement components $D_{1x}, D_{1y}, D_{2x}, D_{2y}$ in the generalized model of energy-dependent susceptibility (1.6), the coefficients A and B being arbitrary

$$D_{1x} = \alpha_1\left\{A(W_1 + W_2) + \varepsilon_0\chi A\left(\alpha_2^2 + \frac{b_1 b_2 \alpha_2}{\alpha_1}e^{-2i\theta}\right)\right.$$
$$\left. + \varepsilon_0\chi B\left[2\left(\alpha_2^2 + \frac{b_1 b_2 \alpha_2}{\alpha_1}\right) + \alpha_1^2 + b_1^2 e^{-2i\theta}\right]\right\}$$

$$(1.353)$$

$$D_{1y} = b_1\left\{A(W_1 + W_2) + \varepsilon_0\chi A\left(b_2^2 + \frac{\alpha_1 \alpha_2 b_2}{b_1}e^{2i\theta}\right)\right.$$
$$\left. + \varepsilon_0\chi B\left[2\left(b_2^2 + \frac{\alpha_1 \alpha_2 b_2}{b_1}\right) + b_1^2 + \alpha_1^2 e^{2i\theta}\right]\right\}$$

$$(1.354)$$

$$D_{2x} = \alpha_2\left\{A(W_1 + W_2) + \varepsilon_0\chi A\left(\alpha_1^2 + \frac{b_1 b_2 \alpha_1}{\alpha_2}e^{2i\theta}\right)\right.$$
$$\left. + \varepsilon_0\chi B\left[2\left(\alpha_1^2 + \frac{b_1 b_2 \alpha_1}{\alpha_2}\right) + \alpha_2^2 + b_2^2 e^{2i\theta}\right]\right\}$$

$$(1.355)$$

$$D_{2y} = b_2\left\{A(W_1 + W_2) + \varepsilon_0\chi A\left(b_1^2 + \frac{\alpha_1 \alpha_2 b_1}{b_2}e^{-2i\theta}\right)\right.$$
$$\left. + \varepsilon_0\chi B\left[2\left(b_1^2 + \frac{\alpha_1 \alpha_2 b_1}{b_2}\right) + b_2^2 + \alpha_2^2 e^{-2i\theta}\right]\right\}$$

$$(1.356)$$

Here

$$W_{1,2} = \alpha_{1,2}^2 + b_{1,2}^2$$

$$(1.357)$$

The following symmetry properties habitual to Eqns. (1.353)–(1.357) will be useful in the forthcoming calculations:

$$\begin{aligned}
&D_{1x} \rightleftarrows D_{2x}; D_{1y} \rightleftarrows D_{2y}; \alpha_1 \rightleftarrows \alpha_2; b_1 \rightleftarrows b_2; \theta \rightleftarrows -\theta; \\
&D_{1x} \rightleftarrows D_{1y}; D_{2x} \rightleftarrows D_{2y}; \alpha_1 \rightleftarrows b_1; \alpha_2 \rightleftarrows b_2; \theta \rightleftarrows -\theta;
\end{aligned} \tag{1.358}$$

Thus, the electric displacement components are shown to depend from the phase shift θ. To outline the physical consequences of this dependence simplifying the calculations, let us assume, that the components of interacting fields $\vec{E}_1$ and $\vec{E}_2$ (1.352) are equal before their interaction in the non-linear plate N (Fig. 1.39):

$$\alpha_1 = \alpha_2 = \alpha_0; \qquad b_1 = b_2 = b_0; \tag{1.359}$$

Considering the small phase shifts ($|\theta| \ll 1$) we introduce the dimensionless normalized perturbations of polarization components arising due to interaction of the waves:

$$\alpha_1 = \alpha_0(1+x); \quad \alpha_2 = \alpha_0(1+y); \quad b_1 = \alpha_0\mu(1+p); \quad b_2 = \alpha_0\mu(1+q). \tag{1.360}$$

Here

$$\mu = b_0\alpha_0^{-1} \leqslant 1 \tag{1.361}$$

$$|x| \ll 1; \qquad |y| \ll 1; \qquad |p| \ll 1; \qquad |q| \ll 1.$$

Our goal is to determine the dependences of perturbations x, y, p, q upon the phase θ. It is expedient to rewrite the expressions (1.353)–(1.357) in a normalized form using the designations (1.360)–(1.361)

$$\begin{aligned}
D_{1x} &= \alpha_1\varepsilon_0\chi\alpha_0^2[3(1+\mu^2) - 2i\theta\mu^2 + \Delta_{11}]; \\
D_{1y} &= b_1\varepsilon_0\chi\alpha_0^2[3(1+\mu^2) + 2i\theta + \Delta_{12}]; \\
D_{2x} &= \alpha_2\varepsilon_0\chi\alpha_0^2[3(1+\mu^2) + 2i\theta\mu^2 + \Delta_{21}]; \\
D_{2y} &= b_2\varepsilon_0\chi\alpha_0^2[3(1+\mu^2) - 2i\theta + \Delta_{22}];
\end{aligned} \tag{1.362}$$

here the terms Δ_{ij} contain the amplitude perturbations:

$$\begin{aligned}
\Delta_{11} &= Qx + Ry + \mu^2(3+B)p + \mu^2(3-B)q; \\
\Delta_{12} &= Rx + Qy + \mu^2(3-B)p + \mu^2(3+B)q; \\
\Delta_{21} &= (3+B)x + (3-B)y + Tp + Zq; \\
\Delta_{22} &= (3-B)x + (3+B)y + Zp + Tq;
\end{aligned} \tag{1.363}$$

$$\begin{aligned}
Q &= 2 - \mu^2(1+B); \qquad R = 4 + \mu^2(1+B); \\
Z &= 4\mu^2 + 1 + B; \qquad T = 2\mu^2 - (1+B);
\end{aligned} \tag{1.364}$$

To find the unknown amplitude perturbations as the functions of phase shift θ let us substitute the expressions (1.362) into the set of equations (1.345) determining the steady-state polarization components.

Using the dimensionless frequency tuning Δ (1.346) and generalizing the definitions of normalized intensities I_0 and A_0 (1.347) connected with the steady-state field (α_0, b_0) and pumping wave (α, b) respectively

$$I_0 = I_1 + I_2; \qquad I_1 = \frac{\varepsilon_0 \chi \alpha_0^2 L_n \Omega_p}{L\gamma}; \qquad I_2 = \mu^2 I_1 \qquad (1.365)$$

$$A_0 = A_1 A_0; \qquad A_1 = \frac{\varepsilon_0 \chi \alpha^2 L_n \Omega_p \gamma_c^2}{LT\gamma^3}; \qquad A_2 = \frac{\varepsilon_0 \chi b^2 L_n \Omega_p \gamma_c^2}{LT\gamma^3}$$

$$(1.366)$$

we may derive from Eqn. (1.345) the set of linearized equations describing the amplitude perturbations. Consideration of an unperturbed state $(\theta = 0)$ results in expression coinciding obviously with the relevant equation (1.348) for linearly polarized waves. Taking into account the first order terms with respect to phase θ and perturbations x, y, p, q one can obtain at last the set of equations connecting these perturbations with phase θ

$$(Q + N)x + Ry + \mu^2(3 + B)p + \mu^2(3 - B)q = \mu^2\delta$$
$$Rx + (Q + N)y + \mu^2(3 - B)p + \mu^2(3 + B)q = -\mu^2\delta$$
$$(3 + B)x + (3 - B)y + (T + N)p + Zq = -\delta$$
$$(3 - B)x + (3 + B)y + Zp + (T + N)q = \delta$$

$$(1.367)$$

Coefficients Q, R, Z, T are given in (1.364)

$$N = \frac{1}{\varepsilon_0 \chi \alpha_0^2} \frac{L\gamma}{L_n \Omega_p} (\mathscr{K} + \mathscr{K}^{-1}); \qquad \mathscr{K} = \Delta + 3I_0 \qquad (1.368)$$

parameter δ is proportional to the phase shift θ

$$\delta = 2\mathscr{K}^{-1}\theta \qquad (1.369)$$

The solution of the system (1.367) may be written as

$$x = -\mu^2 p; \qquad y = \mu^2 p; \qquad q = -p;$$

$$p = F\theta; \qquad F = \frac{2I_0}{1 + \mu^2} \{1 + (\Delta + 3I_0)[\Delta + I_0(1 - 2B)]\}^{-1}$$

$$(1.370)$$

Thus, we obtained the perturbations of polarization components stimulated by the small phase shift θ. These perturbations characterize the turning of polarization ellipses of both interacting waves with respect to fixed (X, Y) axes. Herein the total power of each wave W (1.357) remains constant

$$\Delta W = 2\alpha_0 x + 2b_0 p = 2\alpha_0 (x + \mu^2 p) = 0 \qquad (1.371)$$

Thus the relative decrease of intensity of X-directed component due to influence of phase shift is equal to $2x$ (1.371). Therefore the measurements of this intensity could be used for determination of small phase shift θ by means of Eqn. (1.370). It is worthwhile to outline some merits of this method:

1. The measurement of intensity perturbation $2x$ may be checked due to simultaneous observations of perturbations of X-and Y-polarized light components, which must be equal, according to (1.371), to $2x$ and $-2x$ respectively.

2. The intensity perturbation observed by means of set-up illustrated at Fig. 1.39 is proportional to θ (1.370). The same polarization measurements, performed without this set-up would provide the relative intensity perturbation proportional to θ^2. Such difference illustrates the important role of ring cavity scheme in enhancement of resolution in the range of small θ values.

3. An intriguing possibility of enhancement of phase sensitivity of this wave–wave interaction is based on using of resonant effect. The coefficient of proportionality F between the phase and amplitude perturbations (1.370) tends to infinity under the condition:

$$1 + (\Delta + 3I_0)[\Delta + I_0(1 - 2B)] = 0. \qquad (1.372)$$

Since the value I_0 in (1.372) is real, this resonance is available, if the frequency detuning Δ is large enough

$$\Delta^2 \geqslant \Delta_{\text{th}}^2 = \frac{3(1 - 2B)}{(1 + B)^2} \qquad (1.373)$$

Concerning the sign of detuning Δ, it is shown to be opposite to the sign of non-linear susceptibility, i.e., $\Delta < 0$, if $\chi > 0$, and vice versa.

The enhancement of this energy-dependent resonant polarization effect due to frequency tuning and power modulation is illustrated on Fig. 1.41. The factor of enhancement F could be as large as $10^3 \div 10^4$. This effect cause by non-linear index grating exists for the arbitrary values of parameter B in Eqn (1.372), the detuning being restricted by condition (1.373). Thus, the analysis of interaction of elliptically polarized waves in a ring cavity is shown to advance the new perspectives in using of non-reciproval optical phenomena for precize measurements of small phase shifts.

The hysteretic and resonant phenomena discussed here do exist in the vicinity of a resonator eigenfrequency, and thus such phenomena are strongly sensitive to the spectral range of pumping beams. The

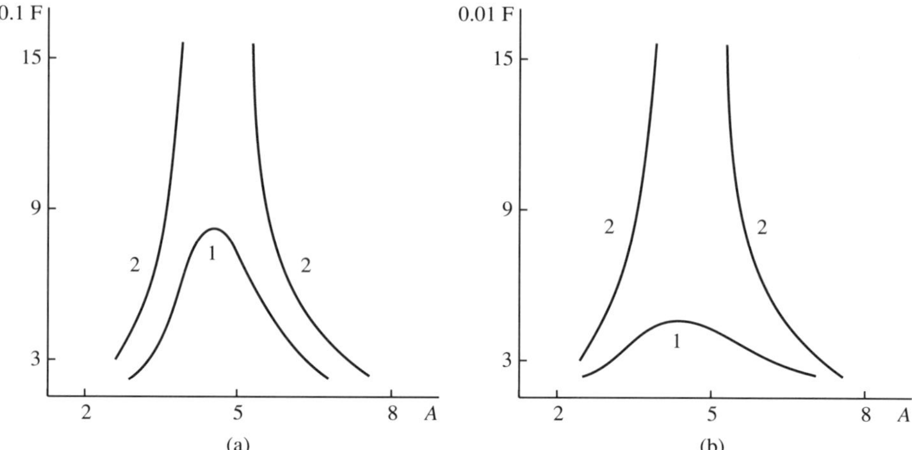

Fig. 1.41   The factor F (1.370) of non-linear resonant enhancement of small phase shifts between the polarization components of elliptically polarized wave ($\mu = 0.5$) is plotted vs the intensity A; (a) the case $B = 0$; curves 1 and 2 relate to the cases $\Delta = -1.6$ and $\Delta = -\sqrt{3}$ respectively; (b) the case B =0.25; the curves 1 and 2 relate to the cases $\Delta = -0.85$ *and* $\Delta = -0.98$ respectively.

wide class of non-resonant hysteresic effects provided by other physical mechanisms of bistability will be discussed in Chapter 2.

CONCLUSION

We conclude with a few comments about the results we have obtained in this chapter.

1. Non-linear cross-modulation of polarization components of the guided mode results in violation of the linear theory predicting an independent propagation of each polarization component with the equal velocities. The linear wave equation describing the mode propagation is transformed due to non-linear consideration to a set of coupled non-linear equations, describing the interaction of polarization components. The energy-dependent difference between the phase velocities of these components leads to the birefringency of waveguide and spatial oscillations of amplitude-phase structure of guided mode (Section 1.4). It is essential, that the period of oscillations depends upon the waveguide geometry; thus, oscillations are vanishing in the square-shaped waveguide. Meanwhile the mode polarization and phases are fixed in

the linear waveguide theory, the non-linear consideration reveals their variability in the course of propagation.

2. The depth of amplitude and phase modulation due to geometric effects is increasing and the spatial period of this modulation is shortening near the waveguide cutoff. Thus the difference of non-linear mode structure from its low intensity limit is deepening close to cutoff.

3. The generalized non-linear versions of twin-core couplers (Section 1.5.1), birefringent waveguides (Section 1.5.2), Bragg filters (Section 1.7.1) exhibit the thresholdless evolution of modes down to the low intensity limits relevant to well-known 'linear' devices. This property permits to outline the peculiar stabilized regimes connected with the mutual compensation of linear and non-linear tendencies. Thus, the condition $T = 0$ (1.257) provides the feasibility of conservation of input polarization in the birefringent waveguide.

4. Analysis of interaction of modes with amplitudes f_1 and f_2 in fibers is shown to be standardized due to systematical using of the auxiliary bilinear functions $f_1 f_2$ and $f_1^2 - f_2^2$ (see, e.g. (1.43)–(1.44), (1.174)–(1.175), (1.224)–(1.225)). This approach reduces the problem to the standard calculations with elliptical functions.

5. Rigorously speaking, the spatial distributions of travelling waves in non-linear waveguide are not the normal modes given, e.g., by (1.67) or (1.116). It is questionable, whether we could use these unperturbed modes in combination with perturbed values of wavenumber β. To obtain the conditions of feasibility of such approximation let us reconsider the non-linear wave equation refusing from the concept of normal modes. Using for simplicity the model of non-linear guiding region surrounded on both sides by dielectric materials we could start with an Eqn. (1.93) for the fundamental TE_{01} mode written in a form:

$$\frac{\partial^2 E}{\partial x^2} + (k_\perp^2 + k_0^2 \varepsilon_0 \chi E^2)E = 0; \qquad (1.374)$$

$k_0 = \omega c^{-1}$ is the vacuum wavenumber. Multiplying Eqn. (1.374) on the factor $2\partial E/\partial x$ we obtain the first integral:

$$\left(\frac{\partial E}{\partial x}\right)^2 + k_\perp^2 E^2 + \frac{k_0^2 \varepsilon_0 \chi E^4}{2} = k_\perp^2 E_0^2 + \frac{k_0^2 \varepsilon_0 \chi E_0^4}{2} \qquad (1.375)$$

Here $E_0 = -E|_{x=0}$ is the peak amplitude of symmetrical mode $E(x) = E(-x)$. Making use of substitution $E^2 = E_0^2 \cos^2 \varphi \, (0 \leqslant \varphi \leqslant \pi/2)$ we can write the solution of Eqn. (1.375) by means of Jacobi elliptical functions cn and sn from the dimensionless coordinate u and modulus p:

$$E = E_0 \mathrm{cn}(u, p). \qquad (1.376)$$

$$u = xk_\perp \sqrt{1 + Q^2}; \qquad p^2 = (2 + Q^{-2})^{-1}; \qquad Q^2 = \frac{k_0^2 \varepsilon_0 \chi E_0^2}{2k_\perp^2}$$

$$(1.377)$$

The field E outside the guiding region may be written in a form

$$E = E_0 \begin{cases} Ae^{-k(x - t/2)}; & x \geqslant t/2; \\ Ae^{k(x + t/2)}; & x \leqslant -t/2; \end{cases} \qquad (1.378)$$

To calculate the parameters A and k one can use the continuity of the field and its first derivative at the boundaries $x = \pm t/2$

$$A = E_0 cn(u_0, p); \qquad u_0 = \frac{k_\perp t}{2} \sqrt{1 + Q^2}; \qquad (1.379)$$

$$k = \frac{k_\perp sn(u_0, p)}{cn(u_0, p)} \sqrt{1 - p^2 sn^2(u_0, p)} \qquad (1.380)$$

Using the definition of parameter k (1.96) determining the depth of field penetration into the regions $|x| > t/2$ we obtain the energy-dependent equation for the transversal wavenamber $k_\perp$

$$\frac{\omega^2}{c^2} (n_0^2 - n_1^2) = k_\perp^2 \left\{ 1 + \frac{sn^2(u_0, p)}{cn^2(u_0, p)} [1 - p^2 sn^2(u_0, p)] \right\};$$

$$(1.381)$$

Unlike the standard approach, utilized everywhere in this Chapter, solution (1.376)–(1.380) describes the field profile $E = E(x, p)$ in the non-linear waveguide without any using of mode concept. The result obtained is a matter of principle, since this self-consistent solution determines simultaneously the unharmonic field profile E and the wavenumber $k_\perp$ both being depended upon the peak amplitude E_0. The low intensity limit of Eqn (1.376)–(1.381)

$$\lim sn(u_0, p)\big|_{p = 0} = \sin u_0; \qquad \lim cn(u_0, p)\big|_{p = 0} = \cos u_0$$

$$(1.382)$$

is coinciding with the equations of linear theory (1.94)–(1.97). The harmonic profile of the mode E in the guiding layer (1.94) in this limit does exist independently of variations of wavenumber. Therefore, using the mode theory for analysis of non-linear waveguides we are dealing with an approximative solutions, the wavenumber perturbations (1.103) being taken into account, meanwhile the mode profile perturbations are ignored (Section 1.3). This approximation is justified for the majority of opto-electronic circuits due to smallness of non-linear parameter Q^2 (1.377): thus the field distribution (1.376) in the case $Q^2 \ll 1$ may be written as a sum of energy-independent function and a small energy-dependent term

$$E = E_0[cos u - \frac{Q^2}{4}(u - sin u \, cos u)]$$ (1.383)

Some other restrictions of applications of mode theory to research of non-linear guiding systems will be discussed at the Conclusion to the Chapter 3.

2.
Non-linear interfaces in optical systems

2.1 Introduction: power-dependent generalization of Fresnel's law

A non-linear interface is an important optical element for spatial and temporal modulation of wave beams. This element consists of the interface between two dielectric materials, one of which has an intensity-dependent refractive index. The interest in non-linear surface wave phenomena is twofold: firstly, these phenomena can provide subsurface localization of powerful wave fields. On the other hand the reflection of light from the electrically-driven interface is an effective tool for tuning the parameters of reflected light waves. The latter effect results in optically controlled tuning of wave refraction also. Puting aside the numerous problems connected with generation of new harmonics at the non-linear boundary, which was the subject of classical research work at the outset of the laser era (Bloembergen *et al.* 1962), we shall analyse the novel possibilities of wave governing beams using power-dependent reflection effects.

To examine the characteristics of non-linear interfaces we must generalize Fresnel's and Snell's laws taking into account the power dependence of the dielectric permittivities of the materials forming the interface. Let us discuss for simplicity, the incidence of S-polarized wave (electric field E_i is orthogonal to the plane of incidence) from the linear medium on to the boundary ($z = 0$) of the non-linear dielectric. (Fig. 2.1). Since we shall consider here media with close values of 'linear' dielectric isotropic permittivities $\mathscr{E}_I$ and $\mathscr{E}_{II}$, it is useful to present the dielectric permittivity of non-linear medium $\mathscr{E}_{II}$ in the form

$$\mathscr{E}_{II} = \mathscr{E}_0(1 + \Delta + \chi_1|\mathbf{E}|^2); \qquad \mathscr{E}_I = \mathscr{E}_0 \qquad (2.1)$$

Here Δ is the linear deviation from the first medium, $\chi_1 = \chi \mathscr{E}_0^{-1}$ characterizes the normalized non-linear susceptibility, and $|\mathbf{E}|$ is the amplitude of the transmitted field E_t. Denoting the amplitude reflected wave by E_r and the angle of incidence by θ, one can represent the amplitudes of the three waves E_i, E_t and E_r, involved in the analysis in the form:

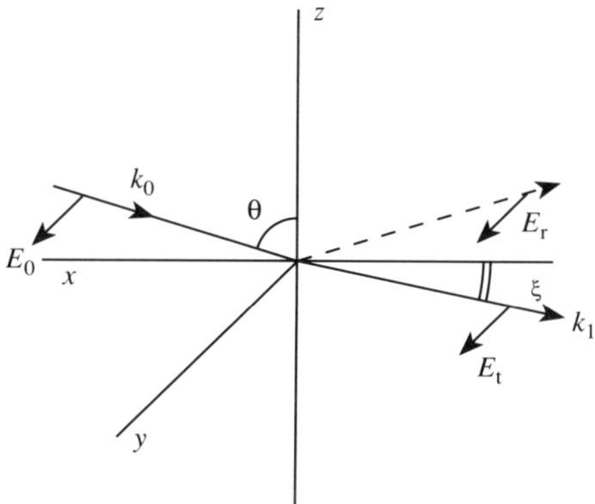

Fig. 2.1 The geometry of a non-linear interface: θ and ξ are the angles of incidence and refraction: $\psi = a/2 - 0$; $\delta = (\pi/2) - \xi$. The incident wave is polarized in the y-direction. The values $z > 0$ relate to the non-linear dielectric.

$$E_i = E_0 \sin(\omega t - k_0 \sin\theta y - k_0 \cos\theta z),$$

$$E_r = RE_0 \sin(\omega t - k_0 \sin\theta y + k_0 \cos\theta z + \alpha), \qquad (2.2)$$

$$E_t = TE_0 F(z) \sin[\omega t - k_0 \sin\theta y + \varphi(z)],$$

where the dimensionless function $F(z)$ describes the transmitted field inside the non-linear medium II. The dissipation of waves in both media is neglected here. R and T are complex reflection and transmission coefficients:

$$R = |R|e^{i\alpha}; \qquad T = |T|e^{i\varphi_0} \qquad (2.3)$$

The boundary values of functions $F(z)$ and $\varphi(z)$, characterizing the transmitted wave, are:

$$F|_{z=0} = 1; \qquad \varphi|_{z=0} = \varphi_0. \qquad (2.4)$$

All the electric vectors $\mathbf{E}_i$, $\mathbf{E}_r$ and $\mathbf{E}_t$ are directed along y-axis (Fig. 2.1). Satisfaction of the boundary conditions at $z = 0$, requiring the continuity of electric field E_y, yields

$$\mathrm{tg}\,\varphi_0 = \frac{|R|\sin\alpha}{1 + |R|\cos\alpha}, \qquad (2.5)$$

$$|T|^2 = 1 + |R|^2 + 2|R|\cos\alpha. \qquad (2.6)$$

This pair of equations expresses the boundary parameters of the transmitted wave through the amplitude and phase of reflection coefficient R. Determining the wave's magnetic field from Maxwell's equation

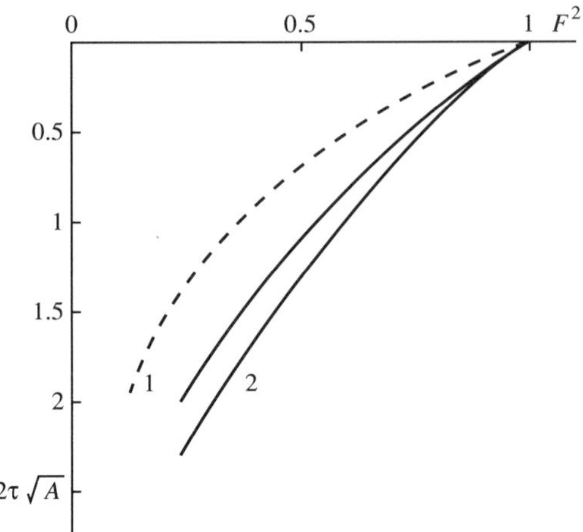

Fig. 2.2 Power-dependent penetration of the incident wave into a non-transparent dielectric. The dimensionless intensity of the refracted field F^2 is plotted against the normalized depth of penetration $2\tau\sqrt{A}$ for several values of the non-linear parameter S(2.32). The broken line and the curves 1 and 2 relate to the linear problem $(S = 0)$, $S = -0.75$, and $S = -0.95$ respectively.

(1.1), imposing the continuity of the components H_z at $z = 0$ and making use of eqn (2.6), we obtain another pair of equations:

$$|R|^2 = 1 + \frac{|T|^2}{k_0\cos\theta}\frac{\partial\varphi}{\partial z}\bigg|_{z=0}, \qquad (2.7)$$

$$\sin\alpha = -\frac{|T|}{2|R|k_0\cos\theta}\frac{\partial F}{\partial z}\bigg|_{z=0}. \qquad (2.8)$$

Both of these pairs are in need in the derivation of generalized Fresnel laws.

These laws depend upon the spatial structure of the transmitted field E_t in medium II. This field obeys the wave equation.

$$\frac{\partial^2 E_t}{\partial z^2} + \frac{\partial^2 E_t}{\partial x^2} + \mu_0\mathscr{E}_{II}\omega^2 E_t = 0. \qquad (2.9)$$

Substituting the dielectric susceptibility (2.1) into eqn (2.9) and using the representation (2.2) of wave E_t, we obtain

$$2\frac{\partial F}{\partial z}\frac{\partial\varphi}{\partial z} + F\frac{\partial^2\varphi}{\partial z^2} = 0, \qquad (2.10)$$

$$k_0^2 F(\cos^2\theta + \Delta + \chi_1 E_0^2|T|^2 F^2) + \frac{\partial^2 F}{\partial z^2} - F\left(\frac{\partial\varphi}{\partial z}\right)^2 = 0. \qquad (2.11)$$

Multiplying (2.10) by F and (2.11) by $\partial F/\partial z$ allows us to obtain two integrals:

$$I = F^2 \frac{\partial \varphi}{\partial z},\tag{2.12}$$

$$I_2 = k_0^2 F^2 \left[\cos^2\theta + \Delta + \frac{\chi_1 E_0^2 |T|^2 F^2}{2}\right] + \left(\frac{\partial F}{\partial z}\right)^2 + \frac{I_1^2}{F^2}.\tag{2.13}$$

According to these integrals two basic types of refracted waves can exist in both linear and non-linear cases:

1. The transmission regime ($R < 1$), which relates to travelling wave E_t with with constant amplitude $F = 1$. In this regime we derive from eqns (2.10, 2.11)

$$\varphi = \varphi_0 - kz,\tag{2.14}$$

$$k = k_0 \sqrt{(\cos^2\theta + \Delta + \chi_1 E_0^2 |T|^2)}.\tag{2.15}$$

Since $\partial F/\partial z = 0$ one can conclude from eqns (2.8) and (2.5), that $\sin\alpha = 0$ and $\varphi_0 = 0$, $T = |T|$. Here the situations $\varphi = 0$ and $\varphi = \pi$ are known to relate to the cases $\mathscr{E}_I > \mathscr{E}_{II}$ and $\mathscr{E}_I < \mathscr{E}_{II}$, respectively, and according to eqn (2.6) these values are connected with relations $T = 1 \pm R$ correspondingly. Combination of eqns (2.7) and (2.15) yields the intensity-dependent Fresnel law for the transmission regime (Kaplan 1980).

$$\chi_1 E_0^2 T^4 + \Delta T^2 + 4\cos^2\theta(T - 1) = 0.\tag{2.16}$$

The low-intensity limit ($E_0 \to 0$) of this equation leads readily to the classical Fresnel formula for the refraction coefficient of S-polarized waves:

$$T = \frac{2\cos\theta}{\cos\theta + \sqrt{\cos^2\theta + \Delta}}\tag{2.17}$$

2. The total reflection regime ($|R| = 1$) is characterized by another intensity-dependent Fresnel law. This regime corresponds to the decrease of the transmitted wave in the bulk of the non-linear medium:

$$\frac{\partial F}{\partial z} \leqslant 0.\tag{2.18}$$

The phase of this wave according to eqns (2.7) and (2.5) is invariant:

$$\varphi = \varphi_0 = \frac{\alpha}{2}.\tag{2.19}$$

Substitution of condition $|R| = 1$ into eqn (2.6) yields

$$T = 2 \cos \frac{\alpha}{2}. \tag{2.20}$$

The relations (2.18–2.20), being independent of the wave's power, are valid in both linear and non-linear regimes. However, in contrast to the linear regime, there are some combinations of parameters θ, Δ, and $\chi_1 E_0^2$, which allow the formation of non-zero field in the bulk of the non-linear medium $(\lim F|_{z \to \infty} > 0)$, whereas other combinations of these parameters are responsible for the appearance of zero field far from the interface $(\lim F|_{z \to \infty} = 0)$. In consequence of these possibilities, two different forms of generalized Fresnel law (Brinca and Mendonca 1979) related to the total reflection regime may be written:

(a) Finite field far from the interface $(z \to \infty)$, which can be determined from eqn (2.11):

$$F_\infty = \sqrt{\left(-\frac{\cos^2 \theta + \Delta}{\chi_1 E_0^2 T^2}\right)}. \tag{2.21}$$

Making use of (2.8), (2.20) and (2.21) we obtain the Fresnel formula:

$$\chi_1^2 E_0^4 T^4 + 2\chi_1 E_0^2 (\Delta T^2 + 4\cos^2 \theta) + (\cos^2 \theta + \Delta)^2 = 0. \tag{2.22}$$

(b) Evanescent field in the depth of non-linear dielectric. Keeping in mind the properties of such a field:

$$\lim F|_{z \to \infty} = 0, \qquad \lim \frac{\partial F}{\partial z}\bigg|_{\tau \to \infty} = 0, \tag{2.23}$$

one can conclude from eqn (2.13)

$$\left(\frac{\partial F}{\partial \tau}\right)^2 = F^2 \left(\cos^2 \theta + \Delta + \frac{\chi_1 E_0^2 T^2 F^2}{2}\right). \tag{2.24}$$

Combination of eqn (2.24) and (2.20) with eqn (2.8) results in an another Fresnel law:

$$\chi_1 E_0^2 T^4 + 2\Delta T^2 + 8\cos^2 \theta = 0 \tag{2.25}$$

These three Fresnel laws (2.16), (2.22), and (2.25) determine the variety of refraction–reflection phenomena on the non-linear interface. The saluent feature of interaction of incidenting waves with the non-linear interface proves to be the formation of bistable amplitude, phase and polarization states in both reflected and transmitted fields. This general tendency is habitual to electrically and thermally-driven interfaces. Since the 'defocusing' non-linearities $(\chi_1 < O)$ of the majority of real materials can surpass pronouncedly the 'focusing' ones, the bistable behaviour of media with $\chi_1 < O$, e.g. InSb, is emphasized below. The analysis is centered here on power-dependent effects stimulated by incidenting waves; the surface waves remain beyond of the scope of consideration in this chapter.

The hysteresis jumps from total reflection to transmission regime and from transmission to total internal reflection are considered in Section 2.2. The bistable variations of amplitudes and phases due to reflection from non-linear dissipative interface is analysed.

Since the electro-optically driven bistability effects proved to be sensitive to the angles of incidence, the methods of controlled tuning of these angles by means of flattening and reshaping of wave beam surfaces are discussed in Section 2.3. The schemes of plane diffractive optical elements transforming the phase surfaces are presented.

Unlike the above-mentioned self-action of incidenting beam at the interface the thermic interaction of two waves with different frequencies is shown in Section 2.4 to tune the reflection properties of thin semiconductor layer. The resonant heating of free carriers in semiconductor plasma provides the deep amplitude-phase modulation of reflected waves, this modulation being different for s- and p-polarizations.

The applications of these results to non-stationary effects of pulses shaping due to switching of bistable thermic states of non-linear semiconductor interface are considered at Section 2.5. The magnetic tuning of these bistability phenomena is also considered.

The non-linear interface effects are generalized for the coated interfaces in Section 2.6. The tunable reflection from multilayer boundary and bistable refraction at the interface coated by diffraction grating are analysed.

We can convince ourselves that this scenario does have a close relationship to the problems of optical signal processing, such as switching or binary logic operations. The simplification of bistable devices due to using of non-cavity schemes and an especially low level of power of input signal extend the possibilities of utilization of non-linear interfaces in all-optical circuits.

2.2 Spatial self-modulation of refracted and reflected fields

Refraction and reflection of powerful light by a non-linear interface may be accompanied by formation of a spatial distribution of refracted field and angular inclination of both refracted and reflected rays. These power-dependent effects are considered to be an effective tool for switching of light in non-linear optical systems. Refraction of a plane wave incident on the non-linear interface is characterized by a generalized Snell's law. According to this law, following from the conservation of wave vector component parallel to the interface, the refraction angle ξ (Fig. 2.1) is given by

$$\mathrm{tg}\,\xi = \frac{\sin\theta}{\sqrt{(\cos^2\theta + \Delta + \chi_1 E_0^2 |T|^2)}}.$$ (2.26)

To obtain the transmission coefficient T in eqn (2.26) it is necessary to use the generalized Fresnel formula. The same coefficient is needed in determination of the spatial structure of the transmitted field due to power-dependent phenomena. Such phenomena, observable for both positive and negative non-linearities, are caused by competition of the linear mismatch Δ of susceptibilities of these media (2.1) and the non-linear component $\chi_1 E_0^2 |T|^2$. As is usual for the effects under discussion, the non-linear component being small ($1 \gg \chi_1 E_0^2$), the only conditions under which these phenomena can be observed are the proximity of the contributions to susceptibility Δ and $\chi_1 E_0^2$ and almost grazing incidence $\psi \ll \pi/2 - \theta$ (Fig. 2.1). Thus, using CS_2, where $\chi_1 = 10^{-20}\,\mathrm{SI}$, one may expect, that an electric field $E_0 = 10^6\,\mathrm{V\,cm^{-1}}$ will be sufficient to produce non-linear perturbations rivalling with $|\Delta| = 5 \times 10^{-4}$, $\psi = 2 \times 10^{-2}$. To bring closer the refractive indices of the dielectrics forming the interface, one may utilize, for example, a glass cell containing CS_2. The glass must be choosen to have an index of refraction close to that of CS_2 at room temperature. This combination can provide a difference of indices as small as $|\Delta| < 10^{-3}$.

Some characteristics of the transmitted field can be found directly from eqn (2.11). Introducing the dimensionless variable $\tau = k_0 z$ connected with free-space wavenumber k_0, we may rewrite this equation with validity in the range $\tau \geqslant 0$ in the dimensionless form

$$\frac{\partial^2 F}{\partial \tau^2} = AF + MF^3$$ (2.27)

with

$$A = -(\psi^2 + \Delta); \qquad M = -\chi_1 E_0^2 |T|^2.$$ (2.28)

The spatial distribution of the transmitted field $F(\tau)$ given by eqn (2.27) with boundary condition

$$F|_{\tau=0} = 1$$ (2.29)

are obtained analytically for the homogeneous non-linear dielectric. However, these distributions depend upon the transmission coefficient

T, which must be determined independently using the generalized Fresnel laws. The possible bistability of this coefficient will be shown to provide hysteretic transitions between the states of the transmitted field. These phenomena attract attention to the power-dependent tuning of transmittivity (Section 2.2.1) and reflectivity (Section 2.2.2) of dielectric films.

2.2.1 *Tunable refraction into a non-linear dielectric*

The amplitude of the refracted wave field is the low-intensity limit described by eqn (2.27) with $M = 0$ is known to be either constant ($A < 0$) or exponentially decreased ($A > 0$):

$$F^2 = \exp(-2\tau\sqrt{A}) \tag{2.30}$$

Scaling the depth of the wave's penetration with the product $\tau\sqrt{A}$, one may illustrate the spatial distribution of intensity F^2 by means of one universal graph (Fig. 2.2, the dotted line). The influence of non-linearity of the dielectrics exhibits different tendencies non-transparent or transparent media in the low-intensity limit.

(a) The non-transparent dielectric $A > 0$. Assuming that the field in the bulk of dielectric is vanishing (2.23), one can find in the case $A + M < 0$ the first integral of eqn (2.27):

$$\left(\frac{\partial F}{\partial \tau}\right)^2 = F^2\left(A + \frac{MF^2}{2}\right). \tag{2.31}$$

Introducing an important parameter,

$$S = \frac{M}{2A} = \frac{\chi_1 E_0^2 T^2}{2(\psi^2 + \Delta)}, \tag{2.32}$$

we obtain the solution of eqn (2.31) in the form

$$\frac{\sqrt{1 + SF^2} - 1}{\sqrt{1 + S} - 1}\frac{\sqrt{1 + S} + 1}{\sqrt{1 + SF^2} + 1} = \exp(-2\tau\sqrt{A}). \tag{2.33}$$

The different values of parameter S determined by the intensity of incident wave are shown to increase (decrease) in the case $-1 < S < 0$ ($S > 0$) an effective depth of penetration characterized in the limit $S \to 0$ by linear approximation (Fig. 2.2). The wave's penetration is strongly increasing in the limit $S \to -1$. Here the refraction coefficient T in (2.32) is given by Fresnel's law (2.25). The low-intensity limit ($S \to 0$) of solution (2.33) leads to a well-known exponential law (2.30).

Unlike these monotonic distributions, the appearance of non-

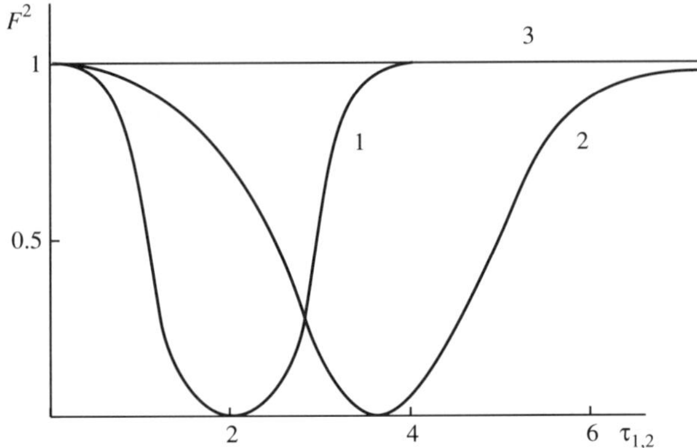

Fig. 2.3 Periodic spatial modulation of intensity F^2 of a powerful refracted wave inside a transparent dielectric. The curves 1, 2, and 3 relate to parameters $A = -1$, $M = 0.5$, $A = 0.3$, $M = -0.75$, and $A = 0.5$, $M = -0.5$, respectively.

harmonic standing field structures proves possible in both transparent and non-transparent dielectrics.

(b) Considering the positive non-linearity $M > 0$, $A + M < 0$, we can write the first integral of eqn (2.27) in the form

$$\frac{2}{M}\left(\frac{\partial F}{\partial \tau}\right)^2 = (1 - F^2)(F_0^2 - F^2).\tag{2.34}$$

Here

$$F_0^2 = -1 - \frac{2A}{M}.\tag{2.35}$$

Recalling the elliptical functions sn and cn used earlier (see Section 1.5.1), we find the solution of eqn (2.34) in the form

$$F^2 = \mathrm{cn}^2(\tau_1); \qquad \tau_1 = \tau\sqrt{-A - \frac{M}{2}}.\tag{2.36}$$

This expression describes the non-linear oscillations of intensity of the refracted wave between the limits $F = 1$ and $F = 0$. The spatial period of this oscillatory structure is (Fig. 2.3)

$$\Lambda_1 = \frac{2\sqrt{2}\,K(k_1)}{\sqrt{-M - 2A}}; \qquad k_1 = \frac{1}{F_0}.\tag{2.37}$$

$K(k)$ is the complete elliptical integral of the first kind; since the parameter F_0 determined by eqn (2.34) obeys in our case $(A < 0, M > 0)$ the condition $F_0^2 > 1$, we obtain that the modulus of the integral $K(k_1)$ is restricted as usual: $k_1 < 1$.

(c) The non-transparent medium, $A > 0$. In contrast to case of moderate non-linearity, $M < 0$, $M + 2A > 0$, discussed above (eqn 2.33), let us consider strong non-linearity, $M < 0$, $M + 2A < 0$. Here the first integral of eqn (2.27) may be written as

$$\frac{2}{|M|}\left(\frac{\partial F}{\partial \tau}\right)^2 = (1 - F^2)(F^2 - F_0^2). \tag{2.38}$$

Proceeding in a similar fashion we obtain the spatial oscillations of intensity between the same limits $F^2 = 1$ and $F = 0$ (Fig. 2.3),

$$F^2 = \mathrm{cn}^2\tau_2, \qquad \tau_2 = \tau\sqrt{|A + M|}, \tag{2.39}$$

distinguished by another period

$$\Lambda_2 = \frac{2K(k_2)}{\sqrt{|A + M|}} ; \qquad k_2 = \sqrt{\frac{|M|}{2|A + M|}} . \tag{2.40}$$

The peculiar case of non-linear field penetration is given by condition $A + M = 0$. The wave's amplitude inside the dielectric in this case remains constant, $F = 1$.

These examples illustrate the possibility of power-dependent tuning of refracted fields in dielectric films. These refraction phenomena will be shown below to exhibit bistable properties.

2.2.2 *Bistable reflection at a non-linear interface*

The reflection of light at the interface between linear and non-linear media can provide the wide class of switching and bistability effects. These effects give rise to possibilities of high-speed tunable power-dependent switching of surface reflectivity from the transmission state to the total reflection regime and vice versa. In contrast to the non-linear Fabry–Perot resonator described in Section 1.8, whose bistability is strongly sensitive to the frequency of incident radiation, the phenomena under discussion are non-resonant and may use a broad spectrum of light.

In deriving the non-linear generalization of Fresnel's law in Section 2.1 we were dealing with an S-polarized incident wave. However, it is worth pointing out that the difference between reflection coefficients $R_{s,p}$ determining the reflection of s- and p- polarized waves from a non-dissipative interface (Fig. 2.1), given by Landau *et al.* (1972) as

$$R_{\rm s} = \frac{\sin(\theta - \xi)}{\sin(\theta + \xi)}, \qquad R_{\rm p} = \frac{{\rm tg}(\theta - \xi)}{{\rm tg}(\theta + \xi)}, \qquad (2.41)$$

is vanishing for glancing angles of incidence and refraction ($\varphi = \pi/2 - \theta$, $\delta = \pi/2 - \xi$, $\varphi \ll \pi/2$; $\delta \ll \pi/2$):

$$R_{\rm s} = R_{\rm p} = \frac{\psi - \delta}{\psi + \delta}. \qquad (2.42)$$

Therefore, the generalized Fresnel laws obtained for grazing incidence are valid for both polarizations of the incident wave beam.

Discussing these laws, we assumed that the incident wave field was described by a plane wave model. That means in this case that the angle of divergence of a diffracted beam of effective radius ρ_0,

$$\psi_{\rm d} = \pi \frac{\lambda_0}{\rho_0}, \qquad (2.43)$$

is much smaller than the critical angle $\psi_{\rm cr}$ of total internal reflection, which is given by relationship

$$\psi_{\rm cr}^2 = |\Delta|. \qquad (2.44)$$

The condition $\psi_{\rm cr} \gg \psi_{\rm d}$ is supposed to be fulfilled in what follows below.

To find the energetic thresholds of hysteretic amplitude–phase switching between transmission and total reflection regimes it is useful to discuss both total reflection (TR) and total internal reflection (TIR)

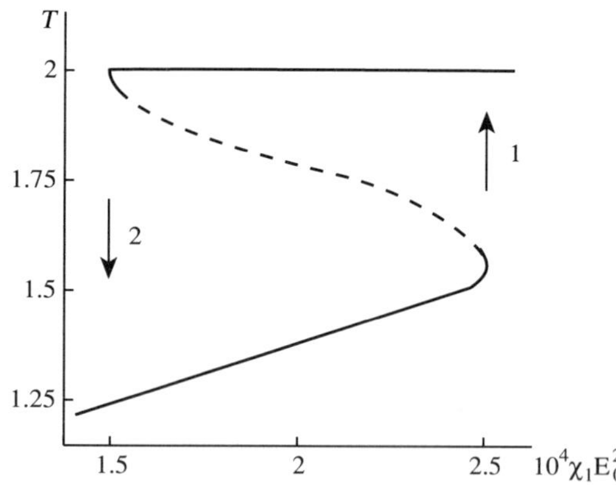

Fig. 2.4 The bistable refraction T of non-linear interface with $\Delta\psi^{-2} = -2.5$ and $\chi_1 < 0$ as a function of input intensity E_0^2. Arrow 1 shows the hysteretic jump between the total reflection and transmission regimes ($E_0 = E_{0\rm c}$) due to increase of intensity E_0^2, while arrow 2 relates to the inverse transition in the course of diminishing of intensity.

regimes, utilizing the idealized model of a non-dissipative interface and a more realistic model taking into account the influence of the wave's dissipation on the bistability of reflection.

(A) The total reflection regime characterized by the condition of non-transparency of the medium bounded by an interface $k^2 < 0$ (2.15) and refraction coefficient $T < 2$ can arise due to reflection from the boundary of the medium with both $\Delta > 0$ and $\Delta < 0$. Considering this regime, e.g. in the case $\Delta < 0$, one can examine the conditions of existence of this regime at such an interface. Choosing the angle of glancing incidence ψ as small as $\psi^2 < |\Delta|$ and, respectively, $A = -(\psi^2 + \Delta) > 0$, we shall discuss according to eqn (2.27) the case of an evanescent field in the bulk of a dielectric in the low-intensity limit. This situation is described by the generalized Fresnel law in the form (1.25), shown in Fig. 2.4. One can see that an increase of incident intensity from zero up to the value E_{0c}^2 determined by eqn (1.25) and the condition

$$\frac{\partial E_0^2}{\partial T} = 0 \tag{2.45}$$

maintains a total reflection regime. Here the refraction coefficient increases from $T_0 = T|_{E_0 = 0}$ up to $T_{cr} = T|_{E_0 = E_{0c}}$:

$$T_0^2 = \frac{4\psi^2}{\Delta} ; \qquad T_c^2 = -\frac{8\psi^2}{\Delta} . \tag{2.46}$$

The jump into a transmission regime accompanied by weakening of reflection appears (Brinca and Mendonca 1979) when the amplitude of the incident wave reaches the values $E_0 = E_{0c}$. With the reduction of E_c imposed, the transmission regime exists until the point

$$T = 2; \qquad \chi_1 E_{01}^2 = -\frac{\psi^2 + \Delta}{4} . \tag{2.47}$$

Since the value of the sum $\psi^2 + \Delta$ in the non-transparent medium is negative, eqn (2.47) yields that the non-linearty of the dielectric under discussion is positive, $\chi_1 > 0$; $CS_2(\chi_1 = 10^{-20}\,SI)$ is an example of such a dielectric. When the reduction of E_0 leads to the value $E_0 = E_{01}$, the jump-like transition returns the transmission to a total reflection regime, completing the hysteretic cycle. E_{0c} and E_{01} are the threshold amplitudes of this cycle.

It is interesting to point out that, unlike the S-like hysteretic graphs (Fig. 1.36), there are no unstable branches in the cycle shown in Fig. (2.4).

(B) Total internal reflection may be considered as an important special case of the above-mentioned total reflection regime characterized by extreme values of the refraction and reflection coefficients $T = 2$, $R = 1$ and refraction angle $\xi = \pi/2$ (Fig. 2.1). The transition of the interface from the transmission regime $A > 0$, $\psi^2 + \Delta > 0$) to total internal reflection available in the case $\Delta < 0$ attracts attention owing to the possibility of using this effect for power-dependent switching of refracted radiation. The energetic thresholds of such a transition can be obtained readily from the generalized Fresnel law in the form (2.16). Small incident amplitudes $E_0 < E_{0c}$, with the threshold value E_{0c} being determined from eqns (2.45) and (2.16), restrict the range of alteration of refraction coefficient T of the transparent dielectric (Fig. 2.5):

$$T\big|_{E_0 = 0} = 2\left[1 + \sqrt{\left(1 + \frac{\Delta}{\psi^2}\right)}\,\right]^{-1}, \tag{2.48}$$

$$T_c = T\big|_{E_0 = E_{0c}} = \frac{3\psi^2}{\Delta}\left[-1 + \sqrt{\left(1 + \frac{8\Delta}{9\omega^2}\right)}\,\right]. \tag{2.49}$$

Substitution of the value $T = T_0$ into Fresnel's law (2.16) leads to a simple connection between the threshold values of intensity E_{0c}^2 and T_c:

$$E_{0c}^2 = \frac{2\psi^2(T_c - 2)}{\chi_1 T_c^4}. \tag{2.50}$$

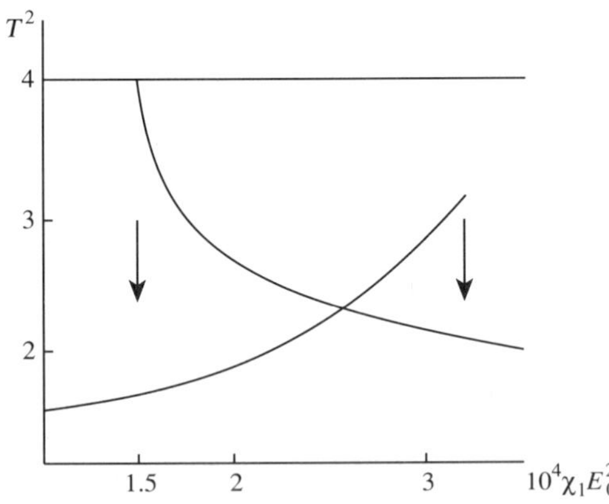

Fig. 2.5 Hysteretic cycles at the non-linear interface with $\Delta\psi^2 = 0.625$ and negative non-linearity, $\chi_1 < 0$. The refraction coefficient T is plotted against the input, intensity E_0^2. The arrows show the jump-like transitions from the transmission to the total internal reflection regimes ($E_0 = E_{0c}$) and vice versa ($E_0 = E_{01}$).

Since $T_c \leqslant 2$, one may conclude that the threshold under discussion exists only at the interface with $\chi_1 < 0$.

The jump-like transition from transmission to total internal reflection show in Fig. 2.5 transforms the system to the state with $T = 2$. The inverse transition to the transmission state appears at the point $E_0 = E_{01}$, which can be found from the condition $\xi = \pi/2$ (2.15):

$$E_{01}^2 = -\frac{\Delta + \psi^2}{4\chi_1}. \tag{2.51}$$

Sometimes it is worth considering the bistability of reflection instead of the above-mentioned bistability of refraction. Using the relation between complex refraction (T) and reflection (R) coefficients,

$$|T|^2 = |1 + R|^2, \tag{2.52}$$

one may examine the hysteretic behaviour of both non-dissipative and dissipative interfaces by means of Fresnel's law for glancing incidence, taking account of both linear attenuation and non-linear absorption derived from generation of new harmonics:

$$R = \frac{\psi - \sqrt{(\psi^2 + \Delta + \chi_1 E_0^2|1 + R|^2)}}{\psi + \sqrt{(\psi^2 + \Delta + \chi_1 E_0^2|1 + R|^2)}}. \tag{2.53}$$

The mismatch Δ and coefficient χ_1 become complex in the dissipative medium, $\Delta = \mathrm{Re}\,\xi + i\,\mathrm{Im}\,\Delta$, $\chi_1 = \mathrm{Re}\,\chi_1 + i\,\mathrm{Im}\,\chi_1$. Rewriting eqn (2.53) in the form

$$1 + R = 2[1 + \sqrt{a + c|1 + R|^2 + i(b + d|1 + R|^2)}\,]^{-1} \tag{2.54}$$

with

$$a = 1 + \frac{\mathrm{Re}\,\Delta}{\psi^2}; \qquad b = \frac{\mathrm{Im}\,\Delta}{\psi^2}; \qquad c = \frac{\mathrm{Re}\,(\chi_1 E_0^2)}{\psi^2}; \qquad d = \frac{\mathrm{Im}\,(\chi_1 E_0^2)}{\psi^2}, \tag{2.55}$$

we can deduce from eqn (2.54) the equations governing the unknown variables $V = |1 + R|^2$ and the phase of complex reflection coefficient $R = |R|\exp{(i\alpha)}$:

$$V^2 = 4[1 + \sqrt{[(a + cV)^2 + (b + dV)^2]} + \sqrt{2}$$
$$\sqrt{\{a + cV + \sqrt{[(a + cV)^2 + (b + dV)^2]}\}}\,]^{-1}, \tag{2.56}$$

$$\cos\alpha = \frac{1}{\sqrt{2}}\sqrt{\left(1 + \frac{a + cV}{\sqrt{[(a + cV)^2 + (b + dV)^2]}}\right)}. \tag{2.57}$$

Amplitude–phase bistabilities of the reflected wave predicted by eqns (2.56, 2.57) are shown in Fig. 2.6.

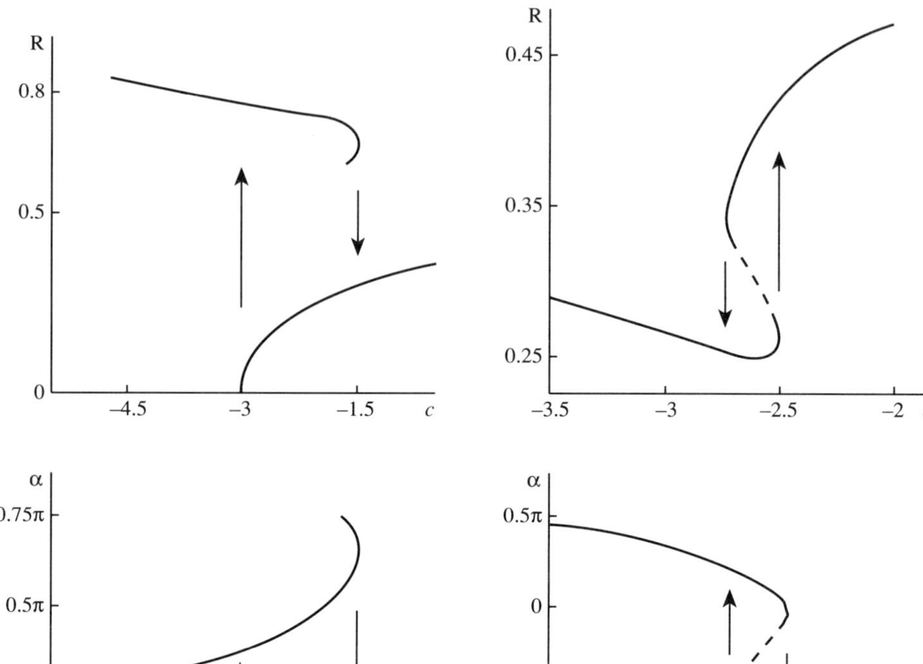

(a)   (b)

Fig. 2.6   Amplitude $|R|$ and phase α of the complex reflection coefficient as functions of input intensity parameters c and d (2.55) for an interface with linear (a) and non linear (b) dissipation; $\psi = 2 \times 10^{-2}$. $\mathrm{Re}\,\Delta = 2 \times 10^{-3}$; $\mathrm{Im}\,\Delta = 2 \times 10^{-4}$; $\mathrm{Im}\,\chi_1 = 0$. (b): $\mathrm{Re}\,\Delta = 3.2 \times 10^{-4}$; $\mathrm{Im}\,\Delta = 16 \times 10^{-4}$; $\mathrm{Re}\,\chi_1 E_0^2 = 0.4 \times 10^{-4}$.

Thus the varying of intensity of incident light at the non-linear interface may provide the series of phenomena useful for amplitude–phase tuning of radiation beams:

1. hysteretic jumps of amplitude and phase of reflected wave;
2. change of penetration depth of the field into the reflecting dielectric;
3. self-limitation of the energy flux of the refracted light due to bistable

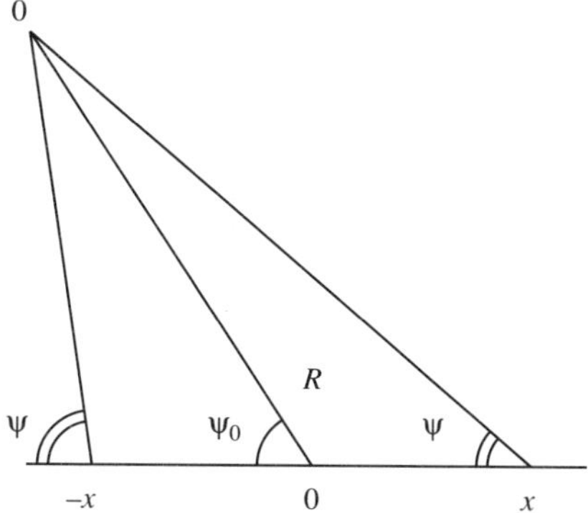

Fig. 2.7 Geometry of illumination of an interface by the glancing axisymmetrical wave beam. The rays' angles of inclination ψ depend upon the position x at the interface. The axial ray passes through the source of radiation 0 and the position $x = 0$.

transitions between transmission and total reflection regimes (Tolminson 1980).

Methods of control of small angles of glancing incidence, which are needed for realization of these possibilities with Kerr-like interfaces, are discussed below in Section 2.3. The broadening of possibilities of hysteretic interface effects based on non-Kerr mechanism of non-linearity and free from the limitations $\psi \ll \pi/2$, $\delta \ll \pi/2$ mentioned in Section 2.2.2 will be discussed further in Section 2.5.

2.3 Plane diffractive optical elements for tuning of amplitude–phase structure of the illuminated area

Use of non-linear interfaces in optical systems is connected with shaping of the illuminated spot and controlled distribution of the wave's amplitude and phase over this spot. The examples discussed above in Sections 2.1 and 2.2 illustrate the significance of small variations of incidence angles of the wave beam for bistable operation with non-linear interfaces. Since these angles are determined by the directions of wave vectors, which are orthogonal to the wave phase surface, the distribution of angles of incidence over the illuminated spot in a realistic model of a pumping beam depends upon the curvature of this phase surface. The related phase-amplitude transformations of the pumping wave beam are accomplished by special plane diffractive optical elements suitable, for example, for change of symmetry or curvature of axisymmetrical phase surfaces of incident wave beams and for reshaping of their intensity profiles (Garitchev *et al.* 1985).

To evaluate the influence of phase front curvature on the energetic characteristics of a non-linear interface, let us recall the strong dependence of threshold parameters E_{0c}^2 (2.50) and E_{01}^2 (2.51), determining the conditions of appearance of bistability, upon the angle ψ of beam incidence. These dependences were obtained in the framework of an idealized model of plane surfaces corresponding to zero curvature. Taking into account the realistic case of non-zero curvature, one may consider the distribution of angles ψ over the illuminated spot (Fig. 2.7). Characterizing some glancing ray coinciding with the beam axis 00 and passing through the point $x = 0$ at the interface at angle ψ_0 we can find, for example, the distribution of angles $\psi(x)$ for other rays passing through the points x belonging to the projection of the aforesaid ray on an interface:

$$
\psi(x) = \begin{cases} \psi_0 \left(1 + \dfrac{x}{R}\right)^{-1}; & x > 0; \\[2em] \left(\psi_0 + \pi \dfrac{|x|}{R}\right)\left(1 + \dfrac{|x|}{R}\right)^{-1}; & x < 0. \end{cases}
\tag{2.58}
$$

Here R is the radius of curvature of the phase surface near the beam axis. The plane wave model ($R \to \infty$) relates to the limit $\psi(x)|_{R \to \infty} = \psi_0$. Substituting the distribution (2.58) into eqn (2.42) we obtain that the reflection coefficient $R(x)$ will change from point to point and, moreover, the zone of existence of bistability will be distributed asymmetrically ($\psi(x) \neq \psi(-x)$; $R(x) \neq R(-x)$) with respect to the beam's centre $x = 0$. This effect restricts the spatial scales of the distability zone.

It is important to point out that the flattening of the phase surface accompanied by the growth of radius of curvature R results, according to eqn (2.58), in diminution of the spread of angles $\psi(x)$ over the illuminated spot. This effect may be useful for tuning of energetic characteristics and spatial scales of non-linear interfaces. Similar results may be provided by transformation of the heterogeneous intensity profile of the pumping beam into a homogeneously illuminated area at the interface. Two examples of plane diffractive optical elements important for non-linear interface operation are discussed below (Jordan *et al.* 1970).

Let us consider the transformation of coherent monochromatic radiation with wavelength λ accomplished by such an optical element located in the (u, v) plane and characterized by a complex transmission coefficient T:

$$T(u, v) = t(u, v)e^{iks(u, v)}. \tag{2.59}$$

Here $t(u, v) = |T(u, v)|$, $k = 2\pi\lambda^{-1}$. Such an element restricted by a non-transparent screen is supposed to be illuminated by a spherical wave

$$E_{\text{in}}(u, v) = |E(u, v)|e^{ikb(u, v)}, \tag{2.60}$$

$$b(u, v) = \sqrt{(R^2 + u^2 + v^2)}. \tag{2.61}$$

The limit $R \to \infty$ relates to the plane wave $E_{\text{in}} = E_0$. The complex amplitude of the pumping wave after its passage through the optical element may be written in the observation plane (x, y) by means of a Kirchhoff–Sommerfeld integral:

$$E(x, y) = Ae^{ikq(x, y)} = \frac{k}{2\pi i} \int E_{\text{in}}(u, v)\, t(u, v)e^{iks(u, v)} \frac{e^{ikl}}{L} \frac{l - z}{L}\, du\, dv, \tag{2.62}$$

$$L = \sqrt{[(x - u)^2 + (y - v)^2 + (l - z)^2]}. \tag{2.63}$$

Here l is the distance between parallel (u, v) and (x, y) planes (Fig. 2.8). The phase surface f of the wave after its passage through the optical element is given by

$$z = f(x, y). \tag{2.64}$$

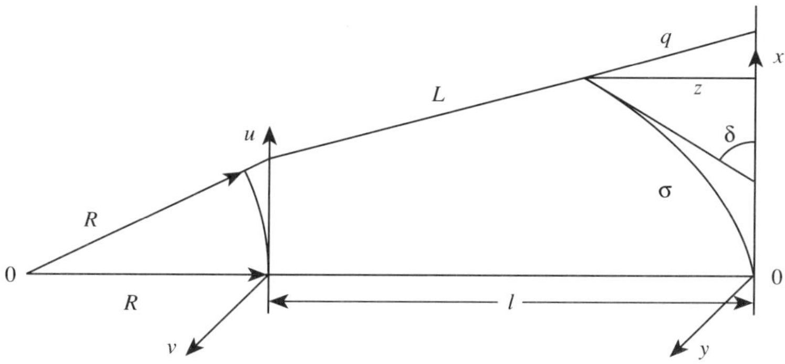

Fig. 2.8 The transformation of a spherical wave irradiated by the source at the position 0 to the wave phase surface f due to passage through the optical diffractive element located in the (u, v) plane. The tangent to the f surface in the (u, v) plane forms the angle δ with the u axis, $\text{tg}\,\delta = \partial f/\partial x$.

The point $z = 0$ relates to the top $(x = 0, y = 0)$ of the phase surface f. The distance q in the phase factor in eqn (2.62) may be calculated from Fig. (2.8):

$$q = |f| \sqrt{\left[1 + \left(\frac{\partial f}{\partial x} \right)^2 + \left(\frac{\partial f}{\partial y} \right)^2 \right]}. \qquad (2.65)$$

Equation (2.62) will be used further for analysis of phase effects stimulated by the two-dimensional profile $s(u, v)$ (2.59).

2.3.1 *Flattening of the wave phase surface*

Substituting the value q (2.65) to a eqn (2.62), we can find the integral equation determining the functions $t(u, v)$ and $s(u, v)$ in the complex transmission coefficient T (2.59). For practical goals it is worth considering a purely phase-diffractive optical element changing the phase of a transmitted wave in the (u, v) plane wherein its amplitude is not violated:

$$t(u, v) = 1. \qquad (2.66)$$

Introducing the phase function φ,

$$\varphi(x, y, u, v) = b(u, v) + s(u, v) + L(x, y, u, v), \qquad (2.67)$$

one may calculate the integral (2.62) be means of the stationary phase method:

$$A e^{ikq(x,y)} = \frac{|E(u, v)|}{L} \frac{l - z}{L} \frac{e^{ik\varphi(x, y, u, v)}}{\sqrt{(\varphi_{uu} \varphi_{vv} - \varphi_{uv}^2)}}. \qquad (2.68)$$

Here

$$\varphi_u = \frac{\partial \varphi}{\partial u} \, ; \qquad \varphi_v = \frac{\partial \varphi}{\partial v} \, ; \qquad \varphi_{uv} = \frac{\partial^2 \varphi}{\partial u \, \partial v} \, . \qquad (2.69)$$

The values u, v in eqn (2.68), relating to observation point (x, y), may be determined from Fig. (2.8):

$$u = x - (l - f) \frac{\partial f}{\partial x} \, , \qquad (2.70)$$

$$v = y - (l - f) \frac{\partial f}{\partial y} \, . \qquad (2.71)$$

The phase of the complex transmission coefficient $s(u, v)$ after the substitution of values u and v into eqn (2.67) may be written as

$$s(u, v) = -b(u, v) - |l - f| \sqrt{\left[1 + \left(\frac{\partial f}{\partial x}\right)^2 + \left(\frac{\partial f}{\partial y}\right)^2 \right]} \qquad (2.72)$$

This important formula describes the distribution of phase shift $s(u, v)$ over the plane diffractive element for each unknown phase surface $f(x, y)$.

The result (2.62) is simplified in the axisymmetrical geometry:

$$x = \rho \cos \alpha; \qquad y = \rho \sin \alpha; \qquad (2.73)$$

$$u = r \cos \beta; \qquad v = r \sin \beta. \qquad (2.74)$$

The substitution of (2.73, 2.74) into (2.70, 2.71) yields

$$\alpha = \beta, \qquad (2.75)$$

$$r = \rho - [l - f(\rho)] \frac{\partial f}{\partial \rho}, \qquad (2.76)$$

$$s(r) = -b(r) - |l - r(\rho)| \sqrt{\left[1 + \left(\frac{\partial f}{\partial \rho}\right)^2 \right]}. \qquad (2.77)$$

To illustrate the utilization of this formula, let us analyse first the simple diffractive element $s(r)$ transforming a spherical wave front into the axisymmetrical paraboloid characterized near the top $z = 0$ by one radius of curvature R_p. Supposing the point-like radiation source to be located at the paraxial centre of curvature of this paraboloid, we obtain

$$b(r) = \sqrt{R_p^2 + r^2}; \qquad f(\rho) = \frac{\rho^2}{2(R_p + l)}. \qquad (2.78)$$

Combination of (2.78) with (2.77) gives the phase distribution providing the transformation under discussion:

$$s(r) = -\sqrt{R_p^2 + r^2} - \left| 1 - \frac{\rho^2}{2(R_p + l)} \right| \sqrt{\left[1 + \frac{\rho^2}{(R_p + l)^2} \right]} \qquad (2.79)$$

To connect the co-ordinates ρ and r it is necessary to substitute an expression determining $f(\rho)$ (2.78) into (2.76) and to solve the obtained cubic equation using the Cardano formula:

$$\rho = \sqrt[3]{[(R_p + l)^2 r + \rho_0^3]} + \sqrt[3]{[(R_p + l)^2 r - \rho_0^3]}, \qquad (2.80)$$

$$\rho_0^3 = (R_p + l)^2 \sqrt{\left(r^2 + \frac{8R_p^3}{27(R_p + l)} \right)}. \qquad (2.81)$$

Calculation of function (2.79) with the dependence $\rho(r)$ (2.80) taken into account completes the search for the phase distribution $s(r)$.

The transformation from a spherical wave front to a non-axisymmetrical phase surface, e.g. to an elliptical paraboloid

$$f(x, y) = \frac{x^2 + cy^2}{2(R + l)}, \qquad (2.82)$$

characterized by two different main radii of curvature at the top point $z = 0$, may be accomplished by analogy. The phase distribution $s(u, v)$ (2.72) may be written as

$$s(u, v) = -\sqrt{(R^2 + u^2 + v^2)} - \left| l - \frac{x^2 + cy^2}{2(R + l)} \right| \sqrt{\left(1 + \frac{x^2 + c^2 y^2}{(R + l)^2} \right)}. \qquad (2.83)$$

Here x and y are the solutions of the system of equations

$$\frac{xR}{(R + l)} \left[1 + \frac{x^2 + cy^2}{2R(R + l)} \right] = u; \qquad (2.84)$$

$$y \left\{ 1 - \frac{c}{l + R} \left[1 - \frac{x^2 + cy^2}{2(l + R)} \right] \right\} = v. \qquad (2.85)$$

Unlike (2.80) this system must be solved numerically.

One may see (Fig. 2.9) that the curvature of the parabolic wave front formed by phase element $s(r)$ may be smaller than the curvature of a spherical wave front propagating in free space, the sources of radiation being located at the same point R_p. These phase surfaces are close only at the neighbourhood of the top point, $\rho = 0$. Such flattening of the phase front leads to approach of the transformed wave to the

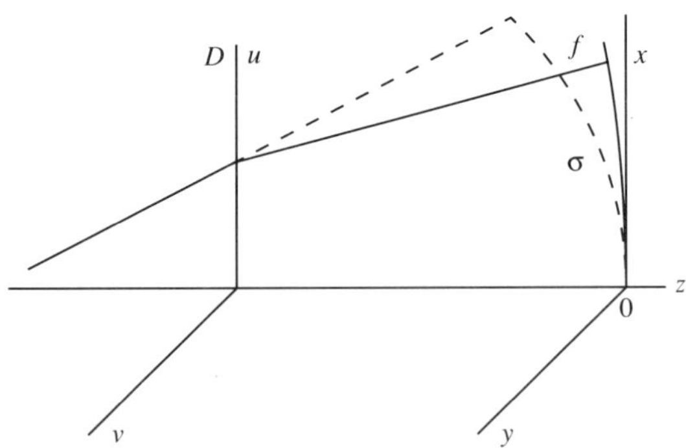

Fig. 2.9　Flattening of the phase front of a spherical pumping wave σ by means of transformation into a paraboloid phase surface f due to passage through plane diffractive optical element D.

plane wave and correspondingly to the narrowing of the distribution of angles $\psi(x)$ (2.58). Moreover, the formation of a non-axisymmetric wave surface f (2.82) provides the appearance of an anisotropic distribution of components of wave vectors $k_\perp$ orthogonal to the axis of the elliptical paraboloid passing through the point $x = y = 0$:

$$\mathbf{k}_\perp = k\nabla_\perp f. \tag{2.86}$$

Here the flattening of wave front may be anisotropic too.

2.3.2 *Reshaping of beam intensity profile*

The formation of the shape of the illuminated area as distinct from the beam's cross section is another problem solved by diffractive optical elements. Herein the condition (2.66) is violated, and obtaining both functions $t(u, v)$ and $s(u, v)$ (2.59) responsible for transformation of a circularly-shaped pumping beam to an illuminated area as a square, ellipse, cross, or set of points important for technological goals is a very complicated problem. Since such figures consist of segments and curve lines, we may simplify the problem by considering the transformation of the initial beam to the small segments and parts of parabolic curves by phase transformations only. Let us for simplicity suppose the pumping wave to be plane.

Using Fig. 2.8 one may write

$$l = L\sqrt{(1 - s_u^2 - s_v^2)}; \qquad s_u = \frac{\partial s}{\partial u}; \qquad s_v = \frac{\partial s}{\partial v}; \tag{2.87}$$

$$x = u + Ls_u; \qquad y = v + Ls_v. \tag{2.88}$$

Eliminating the length L from (2.87–2.88) find

$$x = u + \frac{ls_u}{\sqrt{(1 - s_u^2 - s_v^2)}}, \tag{2.89}$$

$$y = v + \frac{ls_v}{\sqrt{(1 - s_u^2 - s_v^2)}}. \tag{2.90}$$

To solve the above simplified problems, let us represent the function $s(u, v)$ as an expansion

$$s(u, v) = \sum_{p=0}^{p} \sum_{q=0}^{q} s_{pq} u^p v^q. \tag{2.91}$$

In the range of small values ($|u| \ll 1, |v| \ll 1$) the substitution of the phase function

$$s = \frac{v^2 - u^2}{2l} \tag{2.92}$$

into eqns (2.89, 2.90) illustrates the tendency for shaping of the illuminated area at an observation plane in the form of vertical segment

$$x = 0, \qquad y = 2v. \tag{2.93}$$

Now let us consider obtaining the parabolic distribution of intensity at an observation plane:

$$y = \frac{x^2}{2d}. \tag{2.94}$$

Using the expansion

$$\frac{1}{\sqrt{(1 - s_u^2 - s_v^2)}} = 1 + \tfrac{1}{2}\left(s_u^2 + s_v^2\right) - \tfrac{3}{8}\left(s_u^2 + s_v^2\right)^2 + \ldots \tag{2.95}$$

one can find the first terms in the expansion of the phase function (2.91):

$$s(u, v) = -\frac{v}{2}\left[\frac{v}{l} - \frac{d}{2}\left(\frac{u}{l}\right)^2\right]. \tag{2.96}$$

These simple approximate solutions show the rich possibilities for controlled shaping of an irradiated area of complicated form.

2.4 Thermic tuning of the resonant reflectivity of a semiconductor interface

This section is devoted to low-power modulation of the reflectivity of a semiconductor interface for both s- and p- polarizations of pumping wave beam and arbitrary angles of incidence ψ (Fig. 2.10). The well-known method of such modulation is based on growth of free-carrier density in the reflecting layer of semiconductor plasma stimulated by powerful laser radiation of fixed frequency ω responsible for transition of electrons to the conductivity zone; this value is, for example, $\omega = 3.35 \times 10^{14}\,\mathrm{rad\,s^{-1}}$ for InSb. Unlike this rigid condition, thermic modulation effects connected with Joule heating of a collisional semiconductor plasma are not based on the variation of carrier density. This heating can provide effective amplitude–phase perturbations of reflected and transmitted waves in the vicinity of the Langmuir resonance of carriers without any perturbations of carrier density N_e:

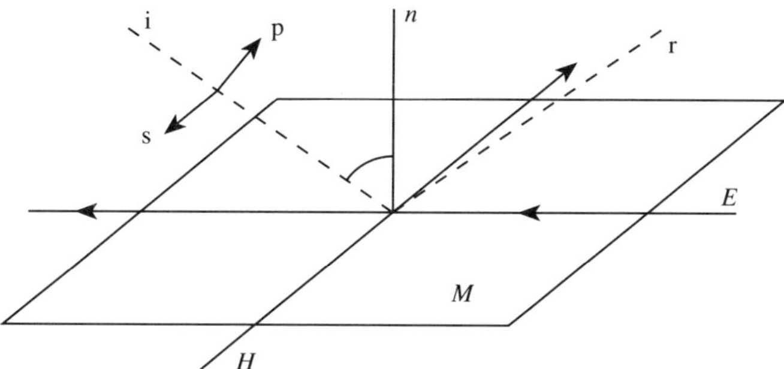

Fig. 2.10 The geometry of tunable modulation of reflected radiation: i and r are the incident and reflected beams, respectively. The symbols s and p illustrate s- and p-polarization of the incident beam. The heating electric field **E** and magnetic field **H** (**H** $\perp$ **E**) are directed parallel to the reflecting film M.

$$\omega = \Omega_{\text{L}}, \tag{2.97}$$

$$\Omega_{\text{L}} = \sqrt{\left(\frac{e^2 N_{\text{e}}}{\mathscr{E}_0 \mathscr{E}_{\text{L}} m^*}\right)}. \tag{2.98}$$

Here e and m^* are the charge and effective mass of free carriers, respectively; $\mathscr{E}_{\text{L}}$ is the dielectric permittivity of the semiconductor crystal lattice; and the vacuum constant $\mathscr{E}_0$ is $8.8 \times 10^{-12}\,\text{F}\,\text{m}^{-1}$.

The influence of heating of carriers, their density being invariant, on the reflectivity of the interface may be analysed by means of standard formulae for the real (Re $\mathscr{E}$) and imaginary (Im $\mathscr{E}$) parts of the dielectric susceptibility of a semiconductor plasma:

$$\text{Re } \mathscr{E} = \mathscr{E}_0 \mathscr{E}_{\text{L}} \left[1 - \frac{\Omega_{\text{L}}^2}{\omega^2 + \nu^2(T_{\text{e}})}\right]; \qquad \text{Im } \mathscr{E} = \frac{\mathscr{E}_0 \mathscr{E}_{\text{L}} \Omega_{\text{L}}^2 \nu(T_{\text{e}})}{\omega[\omega^2 + \nu^2(T_{\text{e}})]}. \tag{2.99}$$

Here $\nu(T_{\text{e}})$ is the frequency of collisions of carriers, dependent on their temperature T_{e}. The heating of carriers results in variations of the temperature-dependent frequency of collisions $\nu(T_{\text{e}})$ and in variations of both the real and imaginary parts of $\mathscr{E}$ (2.99). These thermic perturbations of complex dielectric susceptibility $\mathscr{E}$,

$$\mathscr{E} = \text{Re } \mathscr{E} + i \text{ Im } \mathscr{E}, \tag{2.100}$$

can provide the modulation of complex reflection coefficients for s- and p-polarized waves described by Fresnel's formulae:

$$R_{\mathrm{s}} = |R_{\mathrm{s}}|\,\mathrm{e}^{-\mathrm{i}\varphi_{\mathrm{s}}} = \frac{\cos\psi - \sqrt{(\mathscr{E}\mathscr{E}_0^{-1} - \sin^2\psi)}}{\cos\psi + \sqrt{(\mathscr{E}\mathscr{E}_0^{-1} - \sin^2\psi)}}\;, \qquad (2.101)$$

$$R_{\mathrm{p}} = |R_{\mathrm{p}}|\,\mathrm{e}^{-\mathrm{i}\varphi_{\mathrm{p}}} = \frac{\mathscr{E}\mathscr{E}_0^{-1}\cos\psi - \sqrt{(\mathscr{E}\mathscr{E}_0^{-1} - \sin^2\psi)}}{\mathscr{E}\mathscr{E}_0^{-1}\cos\psi + \sqrt{(\mathscr{E}\mathscr{E}_0^{-1} - \sin^2\psi)}}\;. \qquad (2.102)$$

The separation of real and imaginary parts in (2.102) yields useful expressions for the amplitudes $|R_{\mathrm{s,p}}|$ and phases $\varphi_{\mathrm{s,p}}$ of reflected waves:

$$|R_{\mathrm{s}}|^2 = \frac{\cos^2\psi + a_{\mathscr{E}}^2 - b_{\mathrm{s}}}{\cos^2\psi + a_{\mathscr{E}}^2 + b_{\mathrm{s}}}\;; \qquad (2.103)$$

$$\cos\varphi_{\mathrm{s}} = \frac{\cos^2\psi - a_{\mathscr{E}}^2}{\sqrt{[\,(\cos^2\psi - a_{\mathscr{E}}^2)^2 + b_{\mathrm{s}}^2]}}\;; \qquad (2.104)$$

$$|R_{\mathrm{p}}|^2 = \frac{D_{\mathrm{p}} + a_{\mathscr{E}}^2 - b_{\mathrm{p}}}{D_{\mathrm{p}} + a_{\mathscr{E}}^2 + b_{\mathrm{p}}}\;; \qquad (2.105)$$

$$\cos\varphi_{\mathrm{p}} = \frac{D_{\mathrm{p}} - a_{\mathscr{E}}^2}{\sqrt{[\,(D_{\mathrm{p}} - a_{\mathscr{E}})^2 + b_{\mathrm{p}}^2]}}\;; \qquad (2.106)$$

$$a_{\mathscr{E}}^2 = \left[\mathscr{E}_{\mathrm{L}}\left(1 - \frac{v}{1 + s^2}\right) - \sin^2\psi\right]^2 + \left(\frac{\mathscr{E}_{\mathrm{L}}\,vs}{1 + s^2}\right)\;; \qquad (2.107)$$

$$b_{\mathrm{s}} = 2a_{\mathscr{E}}\cos\psi\cos\theta; \qquad b_{\mathrm{p}} = 2a_{\mathscr{E}}\cos\psi\cos\theta\,(a_{\mathscr{E}}^2 + \sin^2\psi); \qquad (2.108)$$

$$D_{\mathrm{p}} = \cos^2\psi\,(a_{\mathscr{E}}^4 + \sin^4\psi + 2a_{\mathscr{E}}^2\sin^2\psi\cos 2\theta); \qquad (2.109)$$

$$\cos 2\theta = a_{\mathscr{E}}^{-2}\left[\mathscr{E}_{\mathrm{L}}\left(1 - \frac{v}{1 + s^2}\right) - \sin^2\psi\right]. \qquad (2.110)$$

Dimensionless parameters v and s characterize the normalized frequencies of the incident radiation and carrier collisions

$$v = \frac{\Omega_{\mathrm{L}}^2}{\omega^2}\;; \qquad s = \frac{\nu}{\omega} \qquad (2.111)$$

The relaxation of thermic energy of carriers in the low-temperature range is determined mainly by the model of their elastic scattering. The properties of this model, close to those of InSb in the low-temperature range $T_{\mathrm{e}} = 10\,\mathrm{K}$, attract attention because of comparatively strong dependence of the frequency of carrier collisions upon the carrier temperature T_{e} (Zeeger 1973):

$$\nu_i(T_e) = \frac{\nu_{i0}}{f^{3/2}} \; ; \qquad f = \frac{T_e}{T_{e0}} \; . \tag{2.112}$$

Here f is the normalized value of carrier temperature, T_{e0}, and ν_{i0} are the unperturbed values of parameters T_e and ν_i. The temperature dependence $\nu_p = \nu_p(T_e)$ in the case of polar scattering is weaker:

$$\nu_p(T_e) = \frac{\nu_{p0}}{\sqrt{f}} \; . \tag{2.113}$$

Introducing the ratio

$$p = \frac{\nu_{p0}}{\nu_0} \; ; \qquad \nu_0 = \nu_{i0} + \nu_{p0}, \tag{2.114}$$

we may write the temperature dependence of the collisional parameter s (2.111) as

$$s(f) = \frac{s_0}{f^{3/2}} (1 - p + pf); \qquad s_0 = s(f)|_{f=1} \; . \tag{2.115}$$

The characteristic time τ_T of carrier temperature relaxation is known (Mayer and Kellmann 1986) to be evaluated as

$$\tau_T = \frac{kT_{e0}}{m^* \nu_0 u_s^2} \; . \tag{2.116}$$

u_s is the sound velocity in the semiconductor. The light n-type carriers typical, for example, of the $A^{III} B^V$ group can provide values of relaxation time τ_T as small as 1–0.1 ns. Thus, these thermic processes may be considered fast enough.

Substituting the normalized frequency of collisions $s(f)$ (2.115) into Eqns (2.101, 2.102), we may analyse the influence of heating of carriers by means of an external electric field on the complex of reflection and transmission phenomena at the interface of the plasma layer. To point out the possibilities of thermic effects, two examples of using thermically driven interface are discussed further.

2.4.1 *Low-power heating modulation of reflectivity*

The scattering of electrons on ionized impurities, given the strong temperature dependence of frequency of carrier collisions $\nu(T_e)$, is known to dominate in semiconductors of the $A^{III} B^V$ group in the low-temperature range ($T_e \leqslant 10\,\mathrm{K}$). On the other hand, this temperature relates to the low values of equilibrium ionization, N_T, of n-InSb at the temperature $T_e = 10\,\mathrm{K}$. The Langmuir resonant condition (2.97)

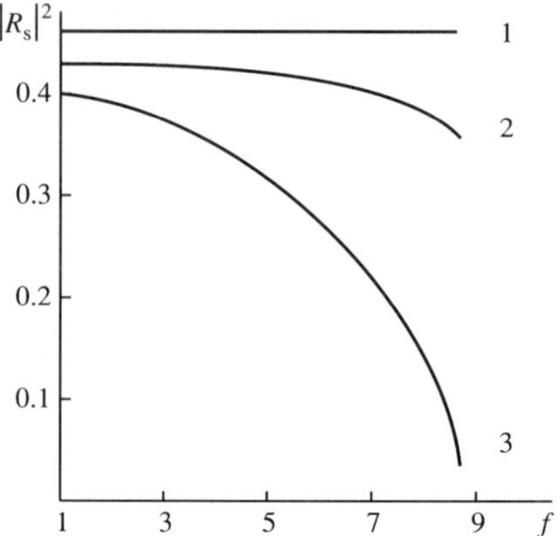

Fig. 2.11   The reflection coefficient $|R_\mathrm{s}|^2$ is plotted against carrier temperature f in n-InSb ($\theta = 45\,°$; $T_{e0} = 78\,\mathrm{K}$). Curves 1, 2, and 3 correspond to the values of parameter $A = N_\mathrm{e}N_\mathrm{L}^{-1} = 0.1,\ 0.5,\ 1$ (2.117), respectively.

permits us to introduce the density of carriers, N_L, connected with the frequency ω of the incident wave:

$$N_\mathrm{L} = \frac{\mathscr{E}_0\mathscr{E}_\mathrm{L}m^*\omega^2}{e^2}.\tag{2.117}$$

These values N_L, which are greater than the equilibrium values N_T, may be reached by alloying of the semiconductor with ionization impurities. The result of alloying may be described by means of the parameter $A = N_\mathrm{e}N_\mathrm{L}^{-1}$. The effectiveness of thermic tuning of the reflection of such an alloying semiconductor grows when the ratio $A = N_\mathrm{e}N_\mathrm{L}^{-1}$ tends to unity (Fig. 2.11). The variation of reflection coefficients depending upon the detuning of the Langmuir resonance for both cases $N_\mathrm{e} \leqslant N_\mathrm{L}$ and $N_\mathrm{e} \geqslant N_\mathrm{L}$ is shown in Fig. 2.12.

The description of the dielectric susceptibility of n-type semiconductors is beyond the scope of quantum theory if the gas of carriers is not degenerate. This means that the density of carriers N_e is limited:

$$N_\mathrm{e} \ll N_\mathrm{cr} = \left(\frac{m^*kT_{e0}}{\hbar^2}\right)^{3/2}\tag{2.118}$$

Here k and $\hbar$ are Boltzmann's and Plank's constants. The limit N_cr for InSb is $N_\mathrm{cr} = 3 \times 10^{17}\,\mathrm{cm}^{-3}$, the temperature being about $300\,\mathrm{K}$; the temperature $T_{e0} = 78\,\mathrm{K}$ for the nitrogen corresponds to the value $N_\mathrm{cr} = 1.5 \times 10^{14}\,\mathrm{cm}^{-3}$. For simplicity we dealing here only with the classical approach (2.118).

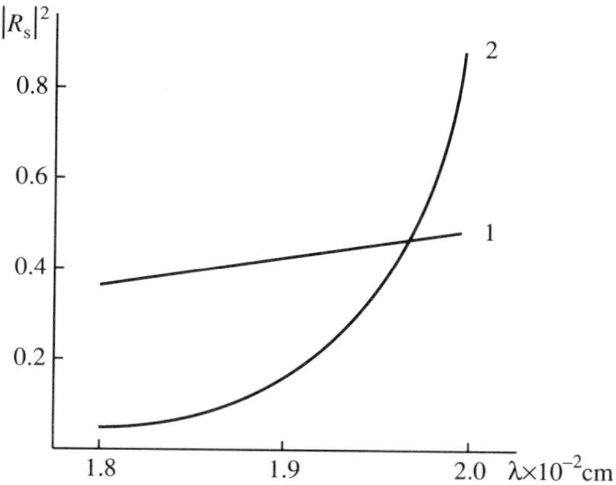

Fig. 2.12 The thermic amplification of spectral filtration properties due to reflection from an n-InSb film near the Langmuir resonance of carriers ($\theta = 45\,^\circ$), $v = 1$). Curves 1 and 2 relate to the values $f = 1$ and $f = 7$, respectively.

A similar resonant effect is manifested in the thermic steepening of the spectral dependence of the reflection coefficient $R(\lambda)$ near the plasma absorption edge (Fig. 2.13). One may note a considerable growth (15–20 times) in the value of the derivative $\partial R^2/\partial\lambda$ in this resonant region.

The energy expenditures for heating of carriers may be found by means of an energy balance equation. Let us consider for simplicity the ohmic heating produced by an external electric field **E** (Fig. 2.10). Since the frequency of collisions in the low-temperature range under discussion is about $v = 10^{11}$ s^{-1} and the frequency of the heating field Ω is often restricted by the condition $\Omega^2 \ll v^2$, we can use the model of plasma heating in a constant field. The heat capacities of the gas of carriers and the crystal lattice in semiconductor different by several orders of magnitude, thus the heating of the lattice is ignored in this problem (Antoniuk *et al.* 1989). One can equate the growth of the thermic energy of the gas of carriers in the plasma and Joule losses due to the conductivity of plasma medium:

$$k v(T_e)(T_e - T_{e0}) = \frac{e^2|\mathbf{E}|^2 \tau_{\mathrm{T}}}{m^* \mathscr{E}_0 \mathscr{E}_{\mathrm{L}} s(T_e)}. \tag{2.119}$$

Here the relaxation time τ_{T} is determined by (2.116), the dimensionless parameter $s(T_e)$ being given by (2.115). The thickness of a layer with hot carriers may be chosen close to the depth of penetration d of the incident wave:

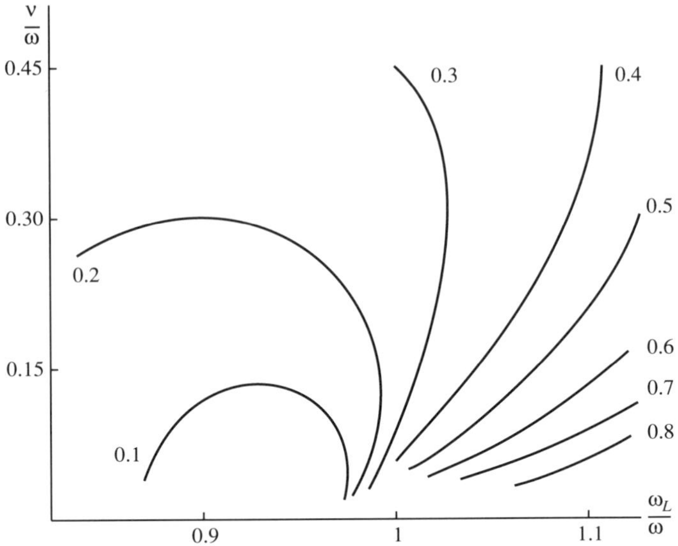

Fig. 2.13   The dependence of reflection coefficient $|R_s|$ ($\theta = 30°$) upon the ratio $v\omega^{-1}$ and the detuning of the Langmuir resonance $d = \sqrt{v} - 1$. The graphs are labelled by the values of $|R_s|$ conserved along the graph; $\theta = 30°$; $\mathscr{E}_L = 15.5$.

$$d = \frac{c}{\omega n_i} \, . \tag{2.120}$$

Here n_i is the imaginary part of the complex refractive index of the semiconductor, $n_r + i n_i$. Determining the temperature of carriers required for a given variation of the reflectivity of the interface, e.g. from Fig. (2.10), and making use of eqn (2.119) we can obtain the power of the heating field per unit of semiconductor volume. Multiplying this by the thickness d of the plasma layer (2.120), we find the expenditures of power per unit of area of interface, Q. Thus, using the value $v_0 = 10^{10}\,\mathrm{s}^{-1}$, $T_{e0} = 10\,\mathrm{K}$, $f = 5$, and $N_e = 5 \times 10^{13}\,\mathrm{cm}^{-3}$, one can find $Q = 10^{-2}\text{–}10^{-3}\,\mathrm{W\,cm}^{-2}$.

To give an idea of thermic tuning of reflectivity we consider here only the steady state of heated carriers. Deferring the analysis of non-stationary heating to Section 2.5, let us discuss first the steady state of transmittivity provided increase of carrier temperature.

2.4.2 *Thermic switching of the transmittivity of thin semiconductor layers*

Along with modulation of the reflectivity of a semiconductor interface, the heating of carriers can provide the variation of thickness d of the skin layer via variation of the imaginary part of the semiconductor refractive index $n_r + in_i$ close to Langmuir resonance. Owing to be exponential dependence of the plasma layer transmittivity upon its thickness z,

$$|T|^2 = (1 - |R|^2)e^{-2z/d}, \tag{2.121}$$

even small heating changes of the parameter n_i may result in effective modulation of transmittivity $|T|^2$. To give a simple example of such modulation, let us consider the case of exact Langmuir resonance ($\omega = \Omega_L, \nu = 1$) and normal incidence of the wave from the air ($\psi = 90°$). Making use of (2.99) we obtain

$$n_r = \sqrt{\left(\frac{\mathscr{L}_L}{2}(D + \sqrt{D})\right)}; \qquad n_i = \sqrt{\left(\frac{\mathscr{L}_L}{2}(-D + \sqrt{D})\right)}, \tag{2.122}$$

$$D = [1 + s^2(f)]^{-1}. \tag{2.123}$$

Diminution of the collisional parameter $s^2(f)$ from 10^{-1} to 10^{-2} owing to growth of carrier temperature by a factor of 2.2 in the model $\nu_i = \nu_{i0}f^{-3/2}$ (2.112) leads to increase of skin layer thickness d by a factor of 3.2 (Fig. 2.14). The alteration of the reflection coefficient $|R|^2$ in this geometry,

$$|R|^2 = \frac{(1 - n_r)^2 + n_i^2}{(1 + n_r)^2 + n_i^2}, \tag{2.124}$$

is insignificant and one may conclude from (2.119) that such heating provides a decrease of transmittivity of $e^{3.2} = 25$ times. Thus, using a methanol laser generating a wavelength of $\lambda_0 = 120\,\mu$m and a semiconductor layer with $z = 200\,\mu$m, we may change the fraction of the

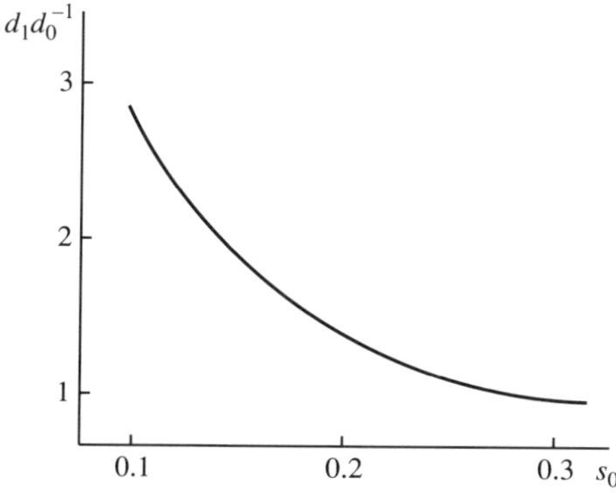

Fig. 2.14 Narrowing of the thickness of the skin layer d for a Langmuir resonant wave ($v = 1$) due to heating of carriers at an n-InSb interface ($s_0 = 0.1, \nu_{p0} = 0, \mathscr{L}_L = 15.5$).

transmitted wave intensity from 0.36 to 0.015.

Langmuir resonant interaction of the incident wave with a heated gas of free carriers may become an effective tool for thermic tuning of both reflectivity and transmittivity of thin semiconductor layers. Such processes are attractive for elaboration of thermic switches characterized by comparatively short delay times ($\tau_T = 10^{-9}$ s), small energy expenditures ($Q = 10^{-2}$–10^{-3} W cm^{-2}), and strong angular dispersion $R^2 = R^2(\psi)$. Modulation of heating power with temporal period $t_0 > \tau_T$ will provide periodic modulation of reflection and transmission coefficients of this thin semiconductor layer. Such element will operate as mirrors or screens with tunable modulation of reflectivity and transparency, respectively, transforming the CW incident wave beam to a sequence of modulated pulses. Use of this effect for shaping of pulses is discussed further.

2.5 Amplitude–phase shaping of pulses due to temporal modulation of the reflectivity of a magnetoplasma layer

The magnetic field $\mathbf{B}$ in a semiconductor plasma is known to induce the birefringency of this medium providing the splitting of a travelling electromagnetic wave into ordinary and extraordinary modes (Zeeger 1973). This magneto-optical effect appears if the frequency of the travelling wave ω becomes comparable with the gyrofrequency of carriers, ω_H:

$$\omega_H = \frac{e|\mathbf{B}|}{m^*} . \tag{2.125}$$

Moreover, the condition stipulating the plasma anisotropy,

$$\omega_H \gg \nu, \tag{2.126}$$

must be fulfilled. The last condition indicates that the magnetic field, which is needed for observation of the well-known Faraday effect, must be strong enough: $|\mathbf{B}| \gg m^*\nu\, e^{-1}$. Thus, in n-InSb we obtain $|\mathbf{B}| = 10^{-1}$–10^{-2} T, the temperature of carriers being about 10 K. Another use of strong magnetic fields is connected with cyclotron resonant heating ($\omega = \omega_H$) of semiconductor plasma carriers (Karlov *et al.* 1985). Unlike the traditional magneto-optical phenomena, we shall discuss the special geometry of the magnetic field, which is unable to perturb directly the refractive index of a layer of collisional semiconductor plasma (Fig. 2.1). However, this magnetic field may change the regime of plasma heating even in the case when the field $\mathbf{B}$ is weak and condition (2.126) is violated. Such an indirect influence of the magnetic field can stimulate the appearance of thermic bistability close to Langmuir resonance and far from the cyclotron resonance. The switching of the magnetic field provides tuning of the reflectivity of the semiconductor interface via the heating of carries. Stationary and non-stationary regimes of resonant magnetothermic modulation of reflectivity of a semiconductor layer are analysed below.

2.5.1 *Thermomagnetic bistability of the resonant reflection of a semiconductor plasma*

To emphasize the role of the magnetic field in the formation of an S-like dependence of carrier temperature upon the heating field $\mathbf{E}$, let us discuss the quasi-stationary limit related to the long period t_0 of the heating field $\mathbf{E} = \mathbf{E}_0 \sin \Omega t$, $\Omega = 2\pi t_0^{-1}$,

$$t_0 \gg \tau_{\mathrm{T}}. \tag{2.127}$$

Here τ_{T} is the temperature relaxation time for carriers (2.116). Generalizing the energy balance equation (2.118) and using the geometry shown in Fig. 2.10, we can calculate the carrier temperature f by means of the dimensionless equation

$$f = 1 + a^2 F(f). \tag{2.128}$$

Here

$$a^2 = \frac{|E_0|^2}{E_{\mathrm{p}}^2}; \qquad E_{\mathrm{p}}^2 = \frac{3kT_{\mathrm{e0}}m^* \nu_0}{e^2 \tau_{\mathrm{T}}}; \tag{2.129}$$

$$F(f) = \frac{1}{2}\left[\frac{1}{(x-y)^2 + s^2(f)} + \frac{1}{(x+y)^2 + s^2(f)}\right]; \tag{2.130}$$

$$x = \frac{\Omega}{\nu_0}; \qquad y = \frac{\omega_{\mathrm{H}}}{\nu_0}. \tag{2.131}$$

The gyrofrequency of carriers, ω_{H}, is determined by (2.125).

Let us consider first the non-magnetized plasma, $\omega_{\mathrm{H}} = 0$, characterized by scattering of carriers on ionized impurities, $s(f) = f^{-3/2}$ (2.112). Equation (2.128) in this approximation may be rewritten in the form

$$(f - 1)\left(x^2 + \frac{1}{f^3}\right) = a^2. \tag{2.132}$$

The dependence of the temperature f upon the energetic parameter a^2 may become bistable in some range of values a^2 found from eqn (2.132) and

$$\frac{\partial a^2}{\partial f} = 0. \tag{2.133}$$

The condition for the rise of bistability is $x < 0.25$, i.e.

$$\Omega < \frac{v_0}{4}. \tag{2.134}$$

The low-temperature state of hysteretic dependence $f = f(a^2)$ related to the beginning of heating is characterized by low-frequency (2.134) absorption of the heating field $n_i v^{-1}$. Subsequently, the decrease of the frequency of collisions during heating leads to formation of a new condition opposite to (2.134), $\Omega > v_i(f)$; here a high-temperature state of carriers characterized by absorption $n_i v(f)[\Omega^2 + v^2(f)]^{-1}$ can arise. The increase of temperature, which is required for transition from the low- to high-temperature state, may be evaluated as $f \geqslant x^{-2/3}$. However, typical values of the above-mentioned parameters for n-InSb are $v_0 = 3 \times 10^{11} \, \text{s}^{-1}$, $\Omega \leqslant 2\pi \times 10^9 \, \text{rad s}^{-1}$; therefore, the increase of carrier temperature required for hysteretic transition must be about $f = 30\text{--}40$; this value is beyond of the scope of the supposed model of elastic scattering of carriers (2.112, 2.113) (Koch *et al.* 1981).

It is important to point out that this trouble can be eliminated by using a magnetic field obeying the condition $\omega_H \gg \Omega$. In this case the variable x in eqn (2.132) must be replaced by y in (2.131) and the condition of appearance of bistability (2.134) must be rewritten in the form $\omega_H < 0.25 v_0$, $y < 0.25$. However, in this case the difference between the values of temperature f related to both temperature states is reduced considerable, and the use of the concept of elastic scattering becomes justified. The case related to magnetic field $|\mathbf{B}| = 5 \times 10^{-3} \text{T}$ $(y = 0.2)$ is shown in Fig. (2.15), curve 1, the polar scattering being ignored $(v_{p0} = 0)$.

It is interesting to emphasize the narrowness of the hysteretic loop $f = f(a^2)$ produced by this thermomagnetic effect: the ratio of the width of the interval of heating field values $|a_1 - a_2|$ responsible for appearance of bistability to the close values $a_{1,2}$ is as small as $|a_1 - a_2| a_{1,2}^{-1} = 2\text{--}5 \times 10^{-2}$.

Examples of bistable dependence $f = f(a^2)$, from scattering on ionized impurities only and by action of both mechanisms of scattering (2.112, 2.113), are shown in Fig. 2.15. One may see, that the presence of polar scattering impedes the hysteretic jumps of temperature and broadens the hysteretic loop. The perturbations of temperature produce, via the dependence $v = v(f)$, the related bistability effects in the amplitude $R(f)$ and phase $\varphi(f)$ of the reflected wave (Fig. 2.16).

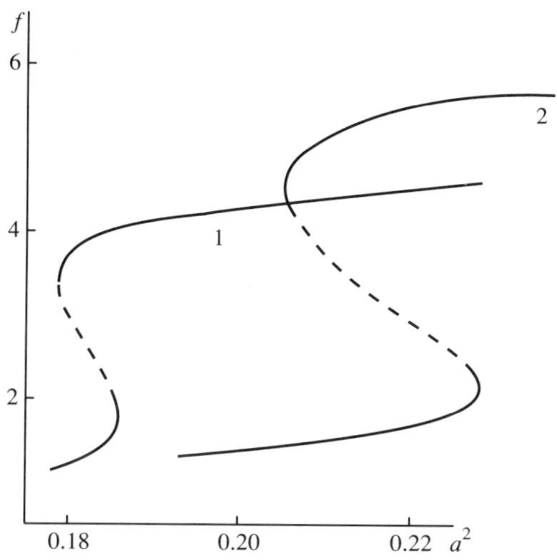

Fig. 2.15 Narrow-loop bistability of the temperature of free carriers in a semiconuctor magnetoplasma heated by an electric field. Curves 1 and 2 relate to the cases $p - 0$, $y^2 - 4 \times 10^{-2}$ and $p = 0.3$, $y^2 = 1.8 \times 10^{-2}$, respectively.

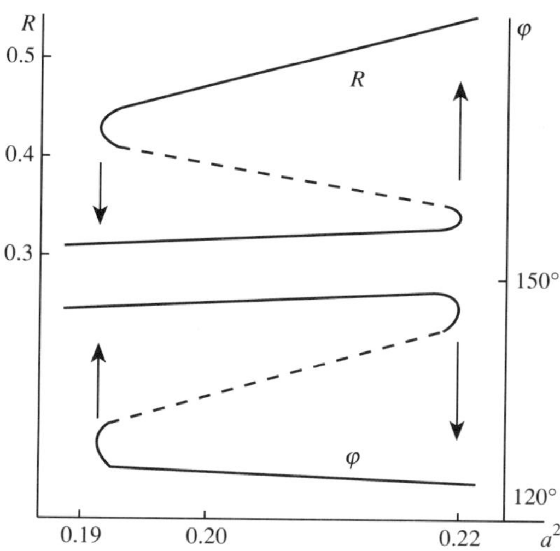

Fig. 2.16 Thermic bistability of amplitude R and phase φ of a reflected S-polarized wave, $\theta = 45\,°$. Variations of the temperature of carriers in the vicinity of Langmuir resonance $(\lambda = 2 \times 10^{-4})$ m; $N = 5 \times 10^{15}\,\mathrm{cm}^{-3}$; $p = 0$; $\nu_{i0} = 10^{11}\,\mathrm{s}^{-1}$) are given by curve 1 of Fig. 2.15.

The aforementioned condition for the appearance of bistability $(y < y_{cr} = 0.25)$ determines the threshold value of the magnetic field B_{th} in this model related to the limit $p = 0$:

$$(B_{th})_0 = \frac{m^* v_0}{4|e|}\,. \qquad (2.135)$$

Here even a small decrease in the magnetic field $B < B_{th}$ will provide considerable growth of carrier temperature via the switching of bistability effect. In contrast to standard graphs illustrating the appearance of bistability in some range of variation of the energetic parameter a^2 (Fig. 2.15), it is interesting to show the appearance of temperature bistability due to variation of the magnetic field, the energetical parameter a^2 being constant (Fig. 2.17). One may see that an S-like temperature dependence $f = f(a)$ appears when the magnetic field is small enough: $y^2 < y_{cr}^2 = 0.25$. Here each value y determines a limited range of values of the energetic parameter $a^2(a_2^2 \leqslant a^2 \leqslant a_1^2)$, $a_1^2 = a_1^2(y); a_2^2 = a_2^2(y)$, corresponding to the bistable regime $f = f(a^2)$. The reduction of the magnetic field leads to reduction of the critical values a_1^2 and a_2^2:

$$\frac{\partial a_1^2}{\partial y} > 0; \qquad \frac{\partial a_2^2}{\partial y} > 0.$$

In order to illustrate magnetic control of the appearance of bistability, let us fix some value $y = y_0 < y_{cr}$ and some value $a^2 = a_0^2$, which obey

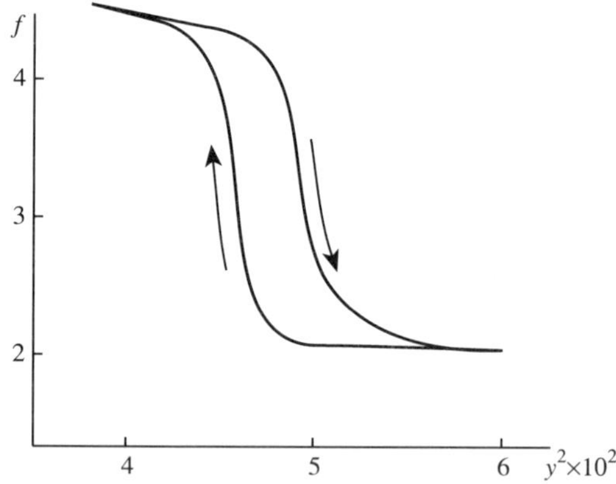

Fig. 2.17 Magnetic tuning of thermic bistability of carriers in a semiconductor plasma. The temperature of carriers f is plotted against the magnetic parameter y^2, the heating electric field being constant ($a_0^2 = 0.175$). The arrows show the variations of temperature f produced by a decrease and an increase of the magnetic field.

the condition $a_2^2(y_0) < a_0^2 < a_1^2(y_0)$. The reduction of the value $y < y_0$, the parameter a_0^2 remaining constant, will produce small variations of temperature values, belonging to the down branch of the S-like curves $f = f(a^2)$ until the condition $a_1^2(y) > a_0^2$ is fulfilled. The reduction of parameter y will introduce the new condition $y < y_1$; $a_1^2(y_1) \leqslant a_0^2$. The appearance of this condition will be accompanied by jump-like transition of values of the related temperature f from the down branch to the upper one (Fig. 2.17). The growth of magnetic field $(y > y_1)$ is accompanied by movement of temperature values f along the upper branches of the S-like curves up to the fulfilment of the condition $y > y_2 : a_0^2 \geqslant a_2^2(y_2)$; at this moment the jump-like transition of temperature to the down branch will appear. The related hysteretic loops for magnetic tuning of amplitude and phase of the reflected wave may be obtained by substitution of temperatures from Fig. 2.17 into the general formulae (2.101, 2.102).

The more realistic model of a semiconductor plasma with both types of collisions (2.112, 2.113) being taken into account may be analysed by means of eqns (2.128) and (2.133). Unlike the scattering on ionized impurities, the polar scattering itself does not lead to bistability effects. Therefore, the growth of parameter $p > 0$, indicating an increase of the contribution of polar scattering, produces a decrease of the threshold value of the magnetic field $(B_{\text{th}})_p$.

$$(B_{\text{th}})_p = (B_{\text{th}})_0 M(p). \tag{2.136}$$

The dimensionless factor $M(p)$ is shown in Fig. (2.18). If the contri-

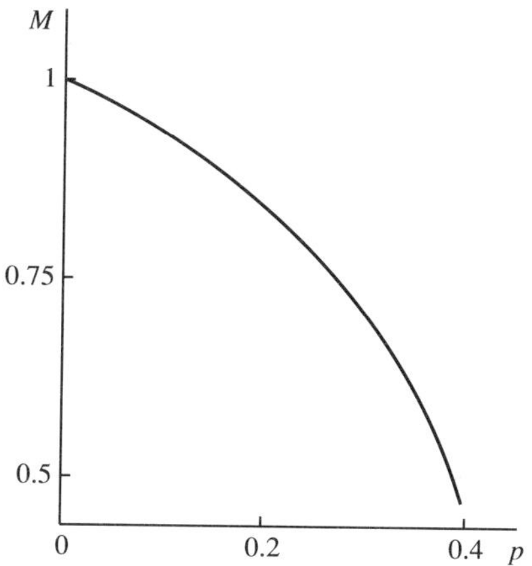

Fig. 2.18 Dependence of magnetic parameter $M(p)$ determining the thermomagnetic bistability upon the types of collisions characterized by parameter p.

bution of polar scattering is sufficient ($p \geqslant \frac{2}{3}$), the appearance of bistability becomes impossible.

Thus the presence of a relatively weak magnetic field provides possibilities of bistable operation with a thermomagnetic interface due to slow switching (2.127) of the heating current or external magnetic field. The high-speed periodic modulation of the heating current with the time scales t_0 and τ_T (2.127) being comparable shows the possibility of shaping of pulses with complicated amplitude–phase profiles.

2.5.2 *Formation of pulses from CW radiation due to modulation of the reflectivity of a magnetoplasma interface*

Periodic modulation of the reflectivity of a semiconductor interface due to heating of carrier magnetoplasma by alternating current, with arbitrary ratio of current period t_0 to relaxation time τ_T, may be analysed by means of a non-stationary energy balance equation:

$$\frac{\partial f}{\partial \eta} = f - 1 + \frac{a^2(\eta)}{y^2 + (1 - p + pf)^2/f^3} . \tag{2.137}$$

Here

$$\eta = \frac{t}{\tau_T} ; \qquad \eta_0 = \frac{t_0}{\tau_T} \tag{2.138}$$

and the energetic parameter a^2 is determined in (2.129).

It is important to point out that harmonic modulation of the heating current (Fig. 2.19a),

$$a(\eta) = a_0 + a_1 \sin\left(2\pi \frac{\eta}{\eta_0}\right), \tag{2.139}$$

can provide non-harmonic oscillations of carrier temperature f due to the non-linearity of eqn (2.137). The depth of temperature modulation is increasing essentially in the range of values of parameter a^2 related to the bistable heating regime (Fig. 2.15). The narrowness of this range, accompanied by noticeable temperature changes, gives rise to obvious asymmetry of the temperature profile $f(\eta)$ (Fig. 2.19b). The temporal modulation of the reflectivity of magnetoplasma layer for a magnetic field of about $|\mathbf{B}| = 2.5 \times 10^{-2}\,\mathrm{T}$ ($\nu_0 = 2 \times 10^{12}\,\mathrm{s}^{-1}$) results in resonant ($v = 1$) transformation of the incident CW wave to an asymmetrically modulated reflected wave; the envelope of modulated signals follows the temperature changes: $R(\eta) = R[f(\eta)]$ (Fig. 2.19).

In contrast, formation of symmetric reflected pulses with the same period $\eta = \eta_0$,

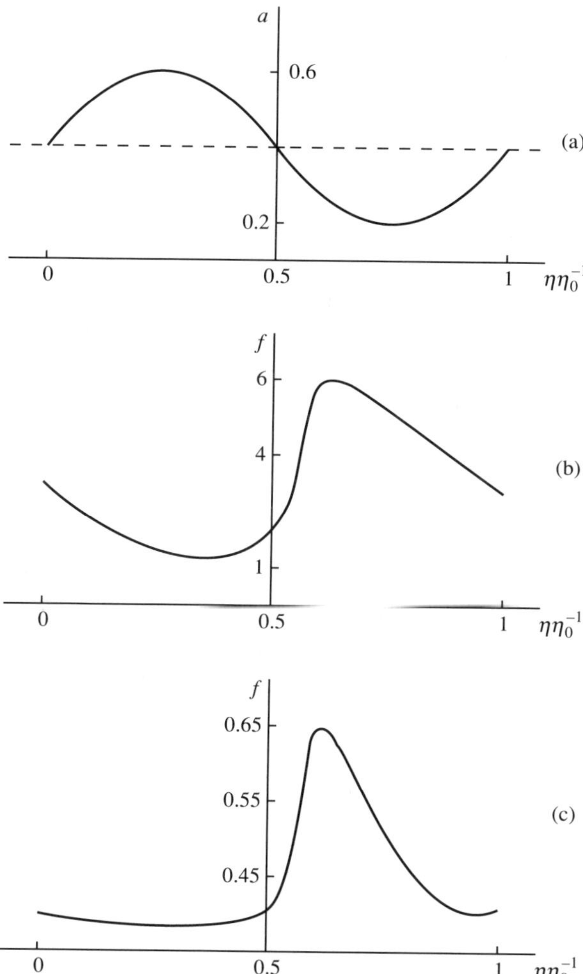

Fig. 2.19 Transformation of an incident s-polarized CW beam ($\theta = 45\,°$) into a series of asymmetrical narrowed wave pulses due to bistable thermomagnetic modulation of reflectivity of interface ($v = 1.08$; $p = 0.25$; $y = 0.14$) by means of symmetrically shaped heating waves $a(\eta/\eta_0)$, $\eta_0 = 30$ (a). (b) and (c) show the temperature profile $f(\eta)$ and amplitude of reflection coefficient $R(\eta)$, respectively.

$$|R| = \left| R_0 - R_1 \cos\left(2\pi \frac{\eta}{\eta_0}\right) \right| \qquad (2.140)$$

is connected with the use of asymmetric heating pulses. The shaping of pulses with maximum amplitude and phase modulation described, respectively, by the ratio $|R_{max}|\,|R_{min}|^{-1} = \tau_{max}$ and the difference $\varphi_{max} - \varphi_{min} = \varphi_m$ is characterized by different combinations of parameters v, s, and ψ. Thus, Fig. 2.20a relates to the deep amplitude modulation $\tau_{max} = 3.2$, the phase modulation being small, $\varphi_m = 13°$, while Fig. 2.20b corresponds to the opposite case of $\tau_{max} = 2$, $\varphi_m = 35°$.

Figure 2.20c relates to s-polarization of the incident wave, whereas the analogous characteristics of the p-polarized wave will be discerned. Therefore, the transformation of an elliptically polarized wave characterized by both s- and p-polarized components may result in modulation of elliptical polarization inside the reflected pulse.

2.6 Coatings for non-linear interfaces

The possibilities of tunable reflectivity of the interface may be broadened by special coatings. Thus the immersion of a non-linear interface into a matching liquid was shown to be a tool for reduction of the difference between their dielectric susceptibilities Δ (2.1) required for the appearance of bistability of the reflected wave. Multi-layer coating of semiconductor interface can increase by several times the variations of reflection coefficients resulting from heating of semiconductor carriers, which was discussed in Section 2.4. Moreover, a weak corrugation of the interface can provide an appearance of bistability for some resonant wavelengths of reflected radiation.

To discuss the simplest properties of non-linear coating it is instructive to compare two problems:

1. Formation of an oscillatary structure of the transmitted field inside the homogeneous medium in the case $A + M > 0$ shown in Fig. 2.3.

2. Penetration of a powerful wave into the heterogeneous layer with a linear profile of dielectric susceptibility

$$\mathscr{E} = \mathscr{E}_0 n^2 \left(1 - \frac{z}{z_0} + \chi|\mathbf{E}|^2\right). \qquad (2.141)$$

Here $n^2 = \mathscr{E}\mathscr{E}_0^{-1}|_{z=0}, E \to 0$. Let us consider for simplicity normal incidence of the wave on the plane $z = 0$. The field structure in the medium (2.141) is described in the range $z > 0$ by the dimensionless equation

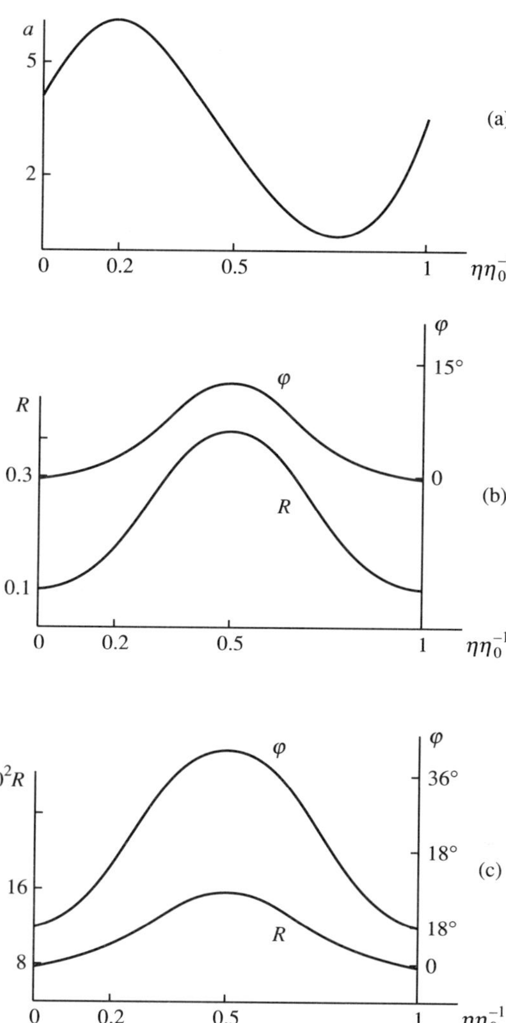

Fig. 2.20 Transformation of an incident s-polarized CW wave ($\theta = 45\,°$) into a series of symmetrically shaped wave pulses due to modulation of reflectivity of the interface ($p = 0.25; y = 0.14$) by means of asymmetrically shaped pulses of heating field $a = a(\eta)$; $\eta_0 = 22.5$ (a). Amplitude and phase profiles of reflected pulses are labelled by letters R and φ for conditions providing either amplitude (b: $v = 0.64, s_0 = 5$) or phase (c: $v = 0.25, s_0 = 2$) modulation of the reflected wave.

$$\frac{\partial^2 a}{\partial \eta^2} + (\eta + a^2)\, a = 0, \qquad (2.142)$$

$$\eta = \left(1 - \frac{z}{z_0}\right)\left(\frac{\omega n z_0}{c}\right)^{2/3}; \qquad a = \sqrt[3]{\left(\frac{\omega z_0}{c n^2}\right)} \sqrt{\chi}\, E. \qquad (2.143)$$

Far from the reflection point, $|\eta| \gg |a|^2$, the contribution of non-linearity is negligible, and the solution of eqn (2.143) in this low-intensity limit is known to be determined by the well-known Airy function (Janke et al 1960). However, close to the reflection point the solution of eqn (2.143) obtained by computer simulation is noticeably different from the limit $a \to 0$ (Fig. 2.21).

The non-linearity of the coating leads to stratification of the transmitted field in a non-transparent region $\eta + a^2 < 0$, unlike the exponential damping of the field typical for the low-intensity limit. Unlike the homogeneous medium (Fig. 2.3), the periods of spatial oscillations of the field as well as their peaks are different, the total amount of these oscillations being increased due to growth of the wave's intensity. Thus, such a coating protects the interface from incident low-intensity radiation.

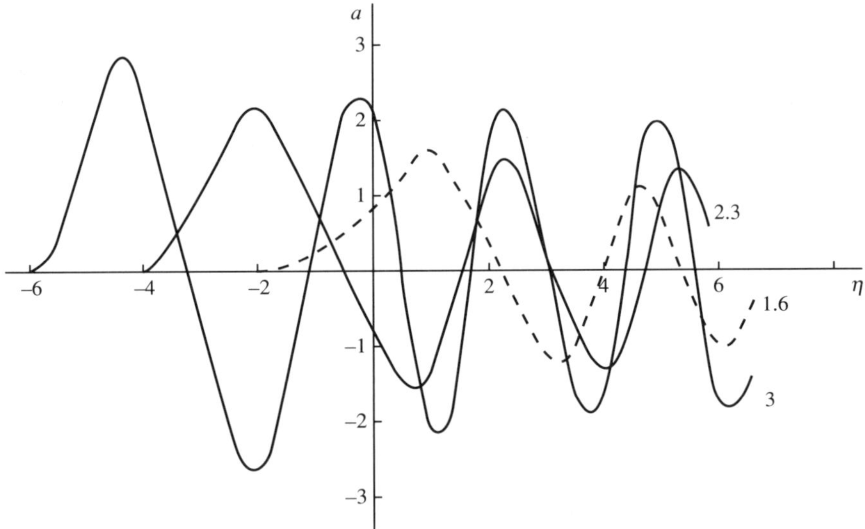

Fig. 2.21 A non-linear standing wave in the heterogeneous transparent layer described by eqn (2.142). The amplitude distributions $a(\eta)$ labelled by the maximum amplitude values are plotted against the dimensionless distance η. The growth of the wave's power leads to formation of new maxima in the non-transparent region $\eta < 0$.

The more complicated models related to coatings of non-linear interfaces and to corrugated non-linear interfaces are discussed below.

2.6.1 *Amplification of the thermic modulation of the wave reflected from a layered structure*

Multi-layer coating of interface is intended for alteration of the amplitide–phase modulation of radiation reflected from the tunable interface. These alterations are based on interference phenomena in

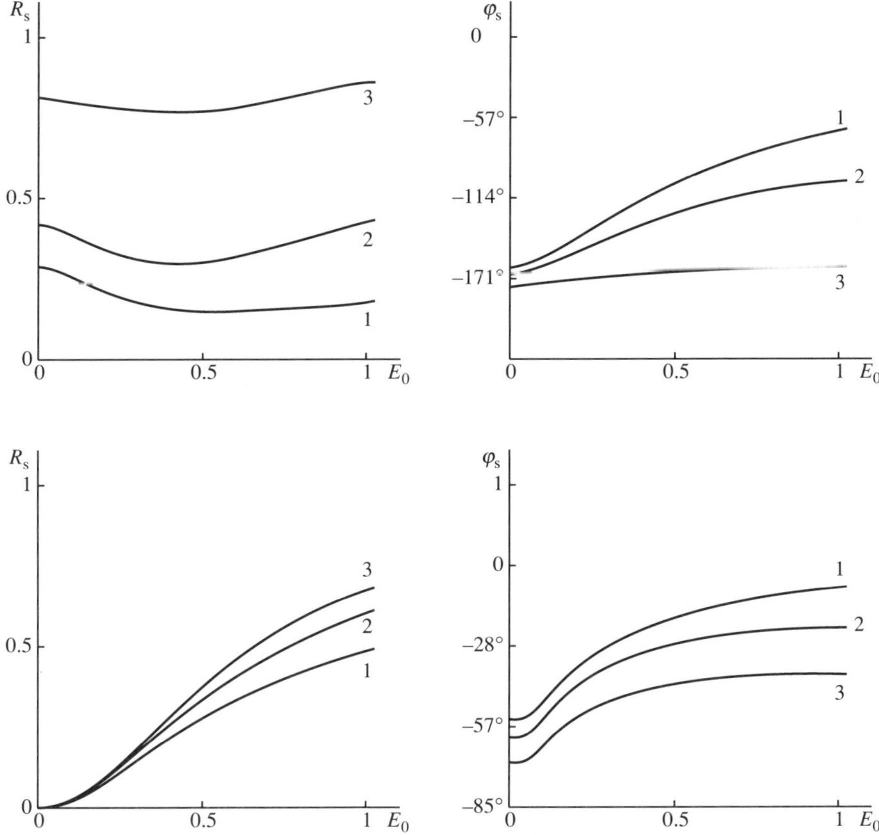

Fig. 2.22 Amplitude R_s and phase φ_s of the reflection coefficient for the reflection of an s-polarized wave from an interface ($v = 1$) without coating (a) and with matching coating (b) are plotted against the voltage of the applied electric field $E_0 E_c^{-1}$; $E_c = 10^5\,\mathrm{V\,m^{-1}}$. Curves 1, 2, and 3 relate to the angles of incidence $\theta = 0$. $\theta = 45\,^\circ$, and $\theta = 80\,^\circ$, respectively. The unperturbed mobility of semiconductor carriers is $\mu = 8 \times 10^3\,\mathrm{cm^2\,V^{-1}\,s^{-1}}$.

layers with different optical lengths. The well-known example of such interference in linear optics is connected with use of $\lambda/4$ plates. Let us consider for simplicity the refractivity of the coating to be independent of the wave's intensity. The complex reflection coefficient of such a coated element may be written as

$$R = \frac{e^{2i\varphi}[R_1(\mathbf{E}) - R_2^*]}{1 - R_1(\mathbf{E})R_2} \ . \tag{2.144}$$

Here $R_1(\mathbf{E})$ and R_2 are the reflection coefficients of the interface and the coating, respectively; φ is the phase shift produced by passage through the coating; and $*$ is the sign of complex conjugation. The reflectivity of such an element in the low-intensity limit $\mathbf{E} \to 0$ tends to zero in the case

$$R_2 = R_1^*\big|_{E=0} \ . \tag{2.145}$$

Condition (2.145) indicates that the reflection may be diminished if the moduli of reflection coefficients of coating and unperturbed interface are equal while their phases have the opposite signs.

A multi-layer coating for each angle of beam incidence φ, for given reflection coefficient R (2.144), may be synthesized by various combinations of layers. The problem of optimization of these combinations of layers is complicated enough, and we shall restrict ourselves here only to illustration of some possibilities of such systems. Using the standard Fresnel formulae for determination of reflection coefficients R_2 taking into account thermic variations of the reflectivity R_1 of the semiconductor interface (Section 2.4), we shall obtain the examples shown in Fig. 2.22. The considerable phase changes of reflected wave obtained due to coating layers may be used in different phase tuning systems.

2.6.2 *Resonant bistability at an interface coated by diffraction grating*

Resonant non-linear interaction of the incident wave with the semiconductor interface can be produced, unlike the Langmuir resonance (Section 2.4), due to the geometric effect of self-action of this wave at the corrugated semiconductor interface. Such self-action may result in the formation of hysteretic reflectivity of the interface; here, in contrast to the effects discussed in Section 2.2, the appearance of bistability is not restricted by glancing angles of incidence of the beam. Such a corrugated interface might be considered here as a diffractive grating covering the non-linear slab (Fig. 2.23). This slab is acting as

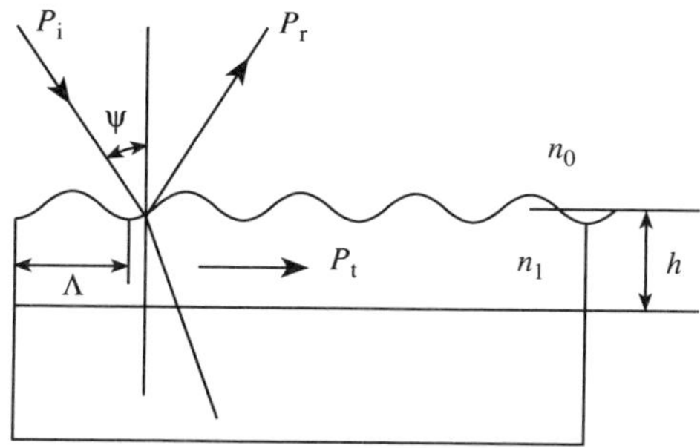

Fig. 2.23 Geometry of reflection of an incident wave beam P_i from the corrugated coating; h, Λ and n_1 are the thickness, period of corrugation, and refractive index of the waveguide formed by this coating.

a corrugated coating possessed with waveguide properties.

The incident s-polarized wave E_i excites the guided mode E_g in the planar slab covered by diffractive grating (Petit 1980):

$$E_g = j_c E_i \left[\left(\frac{\lambda}{\Lambda} + n_0 \sin \psi - n^* \right) - ib \right]^{-1}, \qquad (2.146)$$

$$b = \frac{\gamma_{rad} + \gamma_d}{2k}. \qquad (2.147)$$

Here γ_{rad} and γ_d are the decrements described the radiation and dissipation losses; $\lambda_0 = 2\pi k_0^{-1}$ is the vacuum wavelength of the pumping wave; n^* is the real part of effective refractive index of the guiding system, the effect of corrugation being taken into account; j_c is the dimensionless parameter of coupling, depending upon the geometry of corrugated interface. The non-linear variation of n^* may be expressed via the power of trapped radiation P_t (1.59):

$$n^* = n_0^* + n_2 P_t. \qquad (2.148)$$

Introducing (2.148) into (2.146) we obtain

$$P_t = \frac{j_c^2 P_i}{(\Delta - n_2 P_t)^2 + b^2}. \qquad (2.149)$$

The value j_c can be evaluated by means of the effective thickness h of the corrugated layer (Fig. 2.23) and the dissipative parameter b (2.147): $j_c = 4b(k_0 h)^{-1}$. The detuning parameter Δ is

$$\Delta = \frac{\lambda_0}{\Lambda} + n_1 \sin \psi - n_0^*. \qquad (2.150)$$

From the formal viewpoint eqn (2.149) is close to eqn (1.348). By analogy with eqn (1.348) we can expect the appearance of bistable regimes of reflection described by eqn (2.149). Proceeding in a similar fashion we find the incident powers $P_{i1,2}$ related respectively to switching on and switching off the trapped power (Avrutsky and Sychugov 1990):

$$P_{i1,2} = \frac{2}{27\chi j_c^2} \left[\Delta(\Delta^2 + 9b^2) \pm (\Delta^2 - 3b^2)^{3/2} \right]. \qquad (2.151)$$

It follows from eqn (2.151), that S-like dependence $P_t = P_t(P_i)$ appears if the detuning Δ is large enough:

$$\Delta^2 \geqslant 3b^2. \qquad (2.152)$$

Introducing $\Delta_0 = b\sqrt{3}, \Delta = m\Delta_0 (m \geqslant 1)$, we may rewrite the critical values $P_{i1,2}$ in normalized form using the minimum power P_{i0} required for excitation of the bistable regime:

$$P_{i1,2} = \frac{P_{i1,2}}{P_{i0}} = \frac{m(m^2 + 3) \pm (m^2 - 1)^{3/2}}{4}. \qquad (2.153)$$

The power P_{i0} related to the loseless medium $\gamma \to 0$ is

$$P_{i0} = \frac{8\Delta_0^3}{27\chi j_c^2} \qquad (2.154)$$

Here h is the thickness of the corrugated layer. The corresponded dimensionless values of trapped power $P_{t1,2}$ are

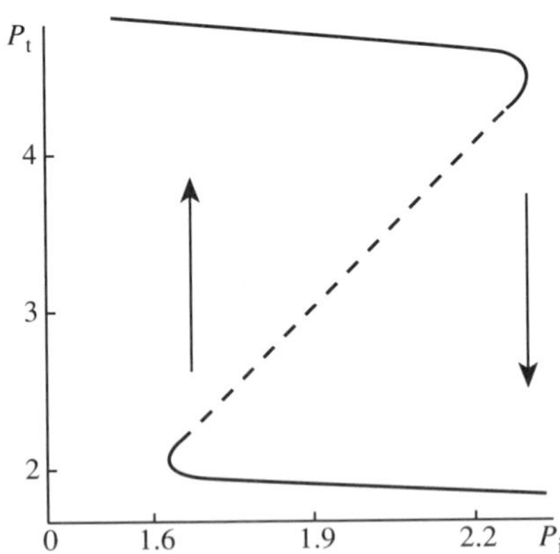

Fig. 2.24 Bistable dependence between the normalized values of power p_t trapped in the corrugated waveguide of Fig. 2.23 with $h = 1\,\mu$m and incident power p_i for the case $m = 1.5$ (2.153), $\Delta_0 = 10^{-6}$, $j_c = 10^{-6}$, $kh = 5$.

$$P_{t1,2} = \frac{P_{t1,2}}{P_{i0}} = \frac{9}{8}\frac{j_c^2}{\Delta_0^2}(2m \pm \sqrt{m^2-1}). \qquad (2.155)$$

The bistable dependence $p_t = p_t(p_i)$ is shown in Fig. 2.24 for some typical values of parameters p_i, Δ, and j_c. The non-linear part of the refractive index χP_t (2.148) in this case is about $\chi P_t = 10^{-6}$. Considering the GaAs semiconductor we obtain an incident power of about $P_i = 1\,\mathrm{MW\,cm^{-2}}$. These jump-like transitions between two levels of trapped power will be accompanied by the similar bistable transitions of reflectivity of the corrugated interface. Along with examples of corrugated fibre sensors (1.271) and couplers (1.303) this example shows again the useful properties of spatially modulated guiding systems.

CONCLUSION

1. Combination of concepts of non-linear interface developed above and guiding system described in Chapter 1 may become useful for elaboration of some optical devices. Thus, the discussion of physical fundamentals of the non-linear interface was restricted here by consideration of the simplest schemes containing one plane boundary between two dielectric media, one of them being non-linear. Extending our knowledge of power-dependent reflection–refraction processes on the surface boundaring the waveguide system, one could analyse the influence of these processes on propagation of radiation in the hollow waveguides produced by drilling of cylindrical channels in some media with the negative part of dielectric permittivity. The diameters of drilled channels are much more, than the trapped optical wavelengths. Specifically, these non-linear surface effects are interesting for waveguide CO_2 lasers using the hollow channels in BeO ceramics.

The bistable variations of wave fields driving due to non-linear interfaces are attractive for operations in all-optical circuits, such as switching or controlled escitement of waveguides. The device using such operations can be constructed in a very small amount of space. Sometimes this design may become compatable with long waveguide spans providing the same amplitude-phase self-modulation by means of gradual evolution of trapped fields.

2. The theory of non-linear interface advanced here is operating with the model of plane homogeneous waves incidenting on this interface. However, the real narrow wave beams using in such devices are characterized by non-zero curvature of wave fronts and power distributions along these fronts, e.g. widely discussed Gaussian beams. These effects restrict the zone of beam–interface non-linear interaction and provide

some new phenomena omitted above. One of these phenomena is connected with Goos–Hanchen shift of the reflected beam along the interface. The picture of this shift is that it is a result of some of the power in the output beam flowing along the interface. Using the ray tracing one may conclude, that the non-linear deepening of rays penetration inside the medium will amplify the displacement of reflected beam axis.

3. It is instructive to outline some problems from this chapter, when the modal theory of non-linear propagation of strong waves proves to be invalid. Such situations appear, if the power-independent and power-dependent changes of refractive index are comparable, thus the theory of non-linear perturbations of mode structure becomes useless. The power-dependent properties of homogeneous non-transparent medium (see Section 2.2) are shown to provide the non-harmonic spatial oscillations of standing wave penetrating inside this medium. As to heterogeneous medium described, e.g. by the linear profile of dielectric permittivity (2.141) the power-dependent field structure is manifested mainly near by the reflection point (Fig. 2.21), where both linear and non-linear variations of refractive index are of the same order of magnitude. The non-linear oscillations of the field in non-transparent region instead of monotonous damping of the field predicted by linear eigenmode, given by well known Airy function, determine the region of uselessness of mode theory in this non-linear problem.

3.
Dispersive non-linear pulse phenomena in guiding circuits

3.1 Introduction: physical foundations of the dynamics of powerful localized waves in dispersive systems

The physical basis non-linear regimes of signal transmission by means of guided systems is the overlapping of some fields of non-linear optics, electrodynamics of bounded systems, and communication theory. The non-linear optical effects point out the physical mechanism of rebuilding of amplitude and spectral envelopes, phase, and the polarization of powerful wave pulses. The mode structure of travelling waves, typical of guided systems, determines the energetic characteristics and spatial scales of such rebuilding. The application of non-linear processes to the coding, transmission and receiving of signals is discussed in the framework of communication theory.

Unlike Chapter 1, this chapter is devoted to the evolution of powerful wave pulses with non-zero spectral bandwidth $\Delta\omega$ in dispersive waveguides. The dispersion leads to dependence of group velocity v_g on the frequency ω. Therefore the pulse duration $2T_0$ will be perturbed by differences between group velocities $v_\mathrm{g}(\omega_1)$ and $v_\mathrm{g}(\omega_2)$ related to different spectral components $E_{\omega 1}$ and $E_{\omega 2}$. The perturbation Δt accumulated after a length L may be ignored, the following condition being fulfilled:

$$\Delta t = L \left| \frac{1}{v_\mathrm{g}(\omega_1)} - \frac{1}{v_\mathrm{g}(\omega_2)} \right| \ll 2T_0. \tag{3.1}$$

Introducing the 'dispersion length' L_ω,

$$L_\omega = \frac{2(v_\mathrm{g} T_0)^2}{|v_\omega|}, \tag{3.2}$$

$$v_\omega = \frac{\partial v_\mathrm{g}}{\partial \omega}, \tag{3.3}$$

we can rewrite condition (3.1) as a restriction of spectral bandwidth required for neglect of dispersive perturbations of the pulse:

$$T_0 |\Delta\omega| \ll \frac{L_\omega}{L}. \tag{3.4}$$

Pulses satisfying this condition ($\Delta\omega \to 0$) were discussed in Chapter 1.

To emphasize the role of non-zero spectral bandwidth in the dynamics of powerful wave pulses, it is worth considering qualitatively the basic effects accompanying this process in the simple model of unbounded Kerr-like non-linear media (1.59). The mutual influence of the phenomena under discussion is disregarded at first. For the wave pulse the intensity is a function time; therefore the refractive index has an instantaneous intensity-dependent part,

$$n(t) = n_0 + n_2(t), \tag{3.5}$$

leading to a phase self-modulation $\varphi(t)$ of a pulse in a length L:

$$\varphi(t) = k_0 L n_2 I(t). \tag{3.6}$$

In order to characterize the influence of this self-modulation on the pulse spectrum we define a non-linear length L_n (1.22):

$$L_\mathrm{n} = \frac{\lambda_0}{4n_2 I}. \tag{3.7}$$

λ_0 is the vacuum wavelength. In a path length L_n the pulse accumulates a maximum phase shift of $\pi/2$ between its peak and periphery. The time-dependent phase shift $\varphi(t)$ is equivalent to a distribution of frequency shift $\Delta\omega(t)$, so-called 'chirp' along the pulse:

$$\Delta\omega(t) = -\frac{\partial\varphi(t)}{\partial t} = -k_0 L n_2 \frac{\partial I(t)}{\partial t}. \tag{3.8}$$

The frequency chirp resulting from the pulse self-phase modulation is seen to be proportional to the derivative of the pulse envelope and is linear only over the central part of the pulse. Note that frequencies in the leading part of the pulse are down-shifted, whereas frequencies in the trailing part are up-shifted (Fig. 3.1). This chirp provides opposite effects in the spectral ranges of normal and anomalous dispersion of refractive index $n(\lambda)$. Thus, in silica fibre the wavelengths $\lambda_0 < 1.3\,\mu$m belong to the range of normal dispersion $v_\omega < 0$ (3.3), while the anomalous dispersion $v_\omega > 0$ is connected with infrared radiation, $\lambda_0 > 1.3\,\mu$m. The chirp is applied to the pulse so that the trailing edge is shifted up in frequency relative to the leading edge. When such a chirped pulse propagates along a dispersive delay line, in the case $v_\omega > 0$ the rear of the pulse will travel faster than the front, the dispersion being anomalous. As a result, the pulse will be compressed in time. In contrast, the non-linear propagation of a pulse in the range $\lambda_0 < 1.3\,\mu$m, e.g. in the visible range, leads to pulse spreading.

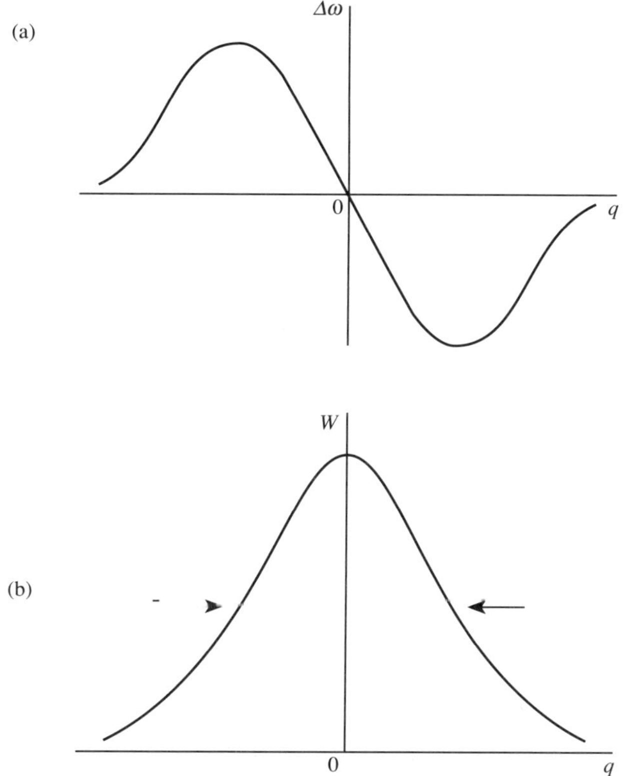

Fig. 3.1 (a) the distribution of dimensionless frequency chirp $\Delta\omega(q)$ $(v(q) = \Delta\omega(q)\,\omega^{-1}$, arbitrary units) along the wave pulse with carrying frequency ω. The arrows in (b) indicate the dispersional perturbations of group velocity $v_1 = v_\omega\,\Delta\omega$ providing the self-confinement of the pulse with the intensity envelope $W(q)$.

If the dispersion parameter v_ω in the spectral range under consideration tends to zero $(v_\omega \to 0)$, the dispersion length is defined by the derivative $v_{\omega\omega}$:

$$L_{\omega\omega} = \frac{6v_0^2 T_0^3}{|v_{\omega\omega}|}; \qquad v_{\omega\omega} = \frac{\partial^2 v_{\mathrm{g}}}{\partial\omega^2}. \tag{3.9}$$

Far from the point $v_\omega = 0$ the ratio $L_{\omega\omega}L_\omega^{-1}$ in the picosecond range is usually as small as $(3\text{–}5) \times 10^{-2}$.

The interaction between non-linear and dispersive distortions of the pulse was not considered in the course of the evaluation of the above-mentioned spatial scales L_ω, $L_{\omega\omega}$, and L_{n}. However, these scales occur naturally in the analysis of competition of amplitude and phase distortions characterizing the evolution of localized wave fields. The basic features of a self-consistent theory of such evolution are described by the non-linear Schrödinger equation. This equation is known to be derivable directly from Maxwell's equations (1.1–1.4) for Kerr-like media (1.6).

The above-mentioned non-linear and dispersive phenomena are intrinsic to both unbounded and bounded media. However, the mode structure of the travelling field in the waveguide complicates the analysis of these problems in guiding systems because the modal dispersion, polarization structure, and heterogeneity of the field over the waveguide's cross-section must be taken into account (Section 1.4). Therefore it is appropriate to start from the simplest form of the non-linear Schrödinger equation (NSE) relating to modes with only one polarization component, such as the TE_{0p} mode. The field distributions over the waveguide's cross-section are assumed to be described by the eigenfunctions of the linear equation, $\psi_{0p}(u)$ (1.67). Substituting such a localized field $\mathbf{E}$ characterized by dimensionless complex envelope $f(z, t)$,

$$E_{0p} = \psi_{0p}(u)f(z, t) \int d\omega\, \mathbf{E}_0(\omega) e^{i[\beta(\omega) - \omega t]}, \qquad (3.10)$$

into Maxwell's equations and expanding the frequency-dependent dielectric permittivity $\mathcal{E}(\omega)$ in a Taylor series up to second order in the carrier frequency ω,

$$\mathcal{E}(\omega) = \mathcal{E}(\omega_0) + (\omega_0 - \omega)\frac{\partial \mathcal{E}}{\partial \omega} + \frac{(\omega_0 - \omega)^2}{2}\frac{\partial^2 \mathcal{E}}{\partial \omega^2}.$$

We shall follow the method of slowly varying amplitudes widely utilized in non-linear field theories (Agrawal 1989). Recognizing that each power of $(\omega_0 - \omega)$ in the summation is equivalent to $i\,\partial/\partial t$ of the summation

$$\mathcal{E}(\omega) = \mathcal{E}(\omega_0) + i\frac{\partial \mathcal{E}}{\partial \omega}\frac{\partial E}{\partial t} - \tfrac{1}{2}\frac{\partial^2 \mathcal{E}}{\partial \omega^2}\frac{\partial^2 E}{\partial t^2},$$

we find the non-linear Schrödinger equation normalized by means of dispersion length L_ω:

$$i\frac{\partial f}{\partial \eta} + \frac{\partial^2 f}{\partial q^2} + \chi_{0p} |f|^2 f = Q_{0p}. \qquad (3.11)$$

Here η and q are the normalized variables

$$\eta = \frac{z}{L_\omega} ; \qquad q = \frac{z - v_g t}{v_g T_0} . \tag{3.12}$$

Parameter χ_{0p} proves to be connected with the characteristic scales L_ω and L_n

$$\chi_{0p} = \frac{\pi}{2} \frac{L_\omega}{L_n} B_{0p} . \tag{3.13}$$

This coefficient B_{0p}, which appears owing to averaging of the heterogeneous field distribution ψ_{0p} over the waveguide's cross-section, may be written formally as a fraction:

$$B_{0p} = \frac{\displaystyle\int_0^\infty |\psi_{0p}(u)|^4 u \, du}{\displaystyle\int_0^\infty |\psi_{0p}(u)|^2 u \, du} . \tag{3.14}$$

As follows from the normalization condition (1.66), the denominator in (3.14) is equal to unity; thus the coefficient B_{0p} is equal to the numerator of the fraction (3.14). In the model of a homogeneous plane wave in an unbounded medium this coefficient tends to unity; for TE_{01} and TE_{02} modes in a circularly shaped waveguide with parabolic profile of refractive index (1.62) these coefficients are

$$B_{01} = \tfrac{1}{4} ; \qquad B_{02} = \tfrac{5}{64} . \tag{3.15}$$

The model $Q = 0$ (3.11) relates to pulse evolution due to the influence of second-order dispersion and non linear perturbation of phase velocity. The non-zero values of Q describe the effects of wave dissipation, power-dependent perturbations of group velocity,

$$v_g = v_{g0}\left(1 - \frac{n_2 I}{n_0}\right), \tag{3.16}$$

and third-order dispersion. Concerning dissipative phenomena, the role of both linear and multi-photon processes will be discussed in Section 3.5. The influence of perturbations of group velocity and third-order dispersion becomes important in when the second-order dispersion tends to zero ($v_\omega \to 0$); the term $\partial^2 f/\partial q^2$ in eqn (3.11) vanishes and use of the scale L_ω (3.2) for determination of the variable η (3.12) and parameter χ (3.13) proves to be impossible. The modified NSE in this limiting case may be written if different dimensionless forms are connected with use of either L_n or $_{\omega\omega}$ lengths for normalization. The choice of related scale depends upon the process under discussion. Thus, the pulse evolution in a dispersionless medium due to power-dependent group velocity is described by an equation normalized by means of the non-linear length L_n:

$$i\frac{\partial f}{\partial \eta} + |f^2|f = -iM\frac{\partial}{\partial q}(|f|^2 f),\tag{3.17}$$

$$\eta = \frac{z}{L_n}; \qquad M = \frac{3 \operatorname{sign} \mathscr{E}_2}{k_0 v T_0}.\tag{3.18}$$

Presenting the function f in (3.17) in the form

$$f = \sqrt{W}e^{i\varphi}, \qquad W = |f|^2,$$

we obtain the set of equations

$$\frac{\partial W}{\partial \eta} + 3MW\frac{\partial W}{\partial q} = 0;\tag{3.19}$$

$$\frac{\partial \varphi}{\partial \eta} + MW\frac{\partial \varphi}{\partial q} = W.\tag{3.20}$$

The solution of eqn (3.19) may be written as

$$W = W(q - 3M\eta W).\tag{3.21}$$

The function W is determined here by the initial conditions; thus, considering the Gaussian initial pulse

$$W|_{\eta = 0} = \exp\left(-\frac{q^2}{q_0^2}\right)\tag{3.22}$$

we find

$$W(\eta) = \exp\left[-\frac{(q - 3\eta MW)^2}{q_0^2}\right].\tag{3.23}$$

Although the peak amplitude of the pulse remains constant in the course of evolution ($W_{max} = 1$), the location of the peak $q_m = 3\eta M$ is shifted towards the leading (trailing) edge in the cases $M > 0$ ($M < 0$), respectively. This shift provides the steepening of one of the pulse's fronts up to formation of electromagnetic shock wave at some point q_c, η_c (Fig. 3.2) determined from the set of equations

$$\frac{\partial W}{\partial q} \to \infty; \qquad \frac{\partial^2 W}{\partial q^2} \to \infty.\tag{3.24}$$

The distance z_c and location of shock wave inside the pulse are given by solutions of eqn (3.24):

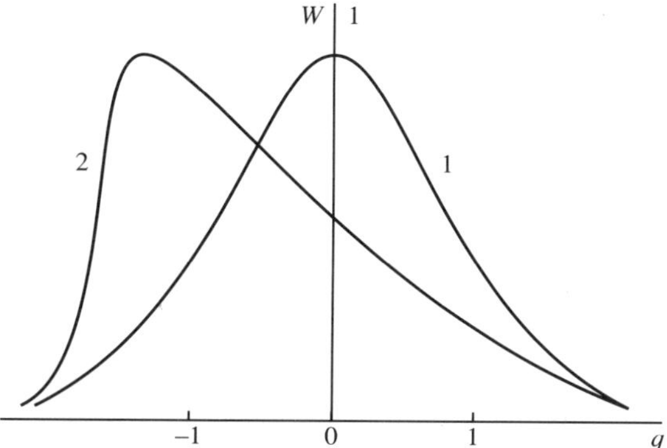

Fig. 3.2 Formation of an electromagnetic shock wave at the trailing edge of an intense Gaussian pulse $W(q)$ in a dispersionless medium with positive non-linearity. Curves 1 and 2 relate to the points $\eta = 0$ and $\eta = 1.15\, z_n$; the spatial scale z_n is given by eqn (3.27).

$$z_c = L_n \eta_c; \qquad \frac{t_c}{T_0} = \frac{z_c}{v_g T_0} - q_c. \qquad (3.25)$$

It is sometimes useful to evaluate the distance z_c approximately without an exact solution of system (3.24). Supposing that the shift of the maximum will provide the formation of a shock wave close to the pulse's edge, we find

$$z_c = \frac{v_g^2 T_0}{|\Delta v_g|}. \qquad (3.26)$$

Substituting Δv_g from (3.16) we obtain the distance of pulse steepening in a dispersionless medium:

$$z_c = \frac{v_g T_0}{(n_2 I)}. \qquad (3.27)$$

The typical example $T_0 = 5\,\text{ps}$, $n_2 I = 10^{-7}$ relates to the distance $z_c = (7\text{--}10)\,\text{km}$. Therefore, such distortions could be accumulated in the long fibres. However, the accumulation of dispersive distortions will impede the shock wave formation due to decrease of pulse shape gradients.

Second-order dispersion in a Kerr-like loseless medium can provide the pulse evolution described by

$$i\frac{\partial f}{\partial \eta_1} - \frac{\partial^3 f}{\partial q^2} + \chi_1 |f|^2 f = 0. \qquad (3.28)$$

Here

$$\eta_1 = \frac{z}{L_{\omega\omega}}; \qquad \chi_1 = \frac{\pi}{2}\frac{L_{\omega\omega}}{L_n}. \qquad (3.29)$$

The dispersion length $L_{\omega\omega}$ is determined by (3.9). The complicated evolution of an initial Gaussian pulse (3.22) described by eqn (3.28) is illustrated in Fig. 3.3. The shift of the central maximum is shown to be accompanied by formation of a new asymmetric maximum.

The latter phenomena may be considered as some correction to the solutions of eqn (3.11) with $Q = 0$. The terms with div **E** vanish from this equation until we are dealing with self-action of TE_{0p} modes containing one polarization component. These modes are used often for analysis of pulse dynamics in this Chapter.

Each of the above-mentioned spatial scales, such as L_ω (3.2), $L_{\omega\omega}$ (3.7), L_n (3.9), z_c (3.27), depending upon the peak amplitude E_0, pulse duration $2T_0$, mode and waveguide characteristics v_g, B_{0p} and n_2, do not depend upon the pulse amplitude-phase envelopes. The dimensionless non-linear Schrödinger equation (3.11) normalized naturally by means of these parameters shows the diversity of non-linear regimes of pulse evolution determined by initial envelops f_0 (q); the scaling lengths L_ω, L_n, $L_{\omega\omega}$ and z_c being the same. The rise and fall of ideas in governed dynamics of non-linear pulses is connected in particular, with the sensitivity of these dynamical processes to initial amplitude-phas distributions.

In extending our knowledge of these processes we may be tempted

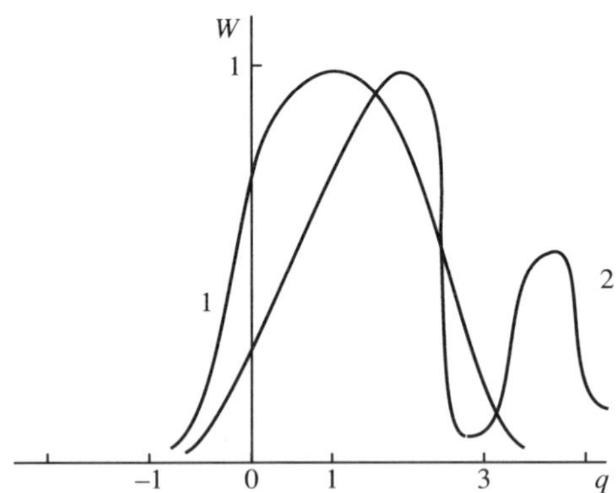

Fig. 3.3 Asymmetric non-linear deformation and self-statification of a Gaussian wave pulse due to second-order dispersion at the distance $z = |L_{\omega\omega}|$. The curves 1 and 2 relate to the values $\chi_1 = 0$ and $\chi_1 = 5$, respectively.

to outline two main tendencies, providing both the acceleration and deceleration of pulse self-action in guiding systems. The first tendency is connected with controlled shaping of solitary waves, meanwhile another one results in stabilization processes. As to the stabilization of specially-shaped pulses important, e.g., for their distortion-free propagation in optical communication channels, this problem has been centered traditionally on research of solitons. Solitons are the pulses, whose shape (sech. intensity envelope) and peak intensity are such, that effects of index non-linearity always exactly cancel dispersive broadening; thus in the limit of zero loss the pulse width and shape remain constant with propagation. Mollenauer *et al.* (1985) presented the experimental evidence of soliton propagation in optical fibers. In view of stability of these pulses they are suggested for using in long communication lines. Since these signals, characterized by fixed invariant shape, have no free parameters and can be used therefore in binary code system only, one pays a penalty for transmission of these stabilized signals due to obvious restriction of information capacity of such lines. To avoid this penalty the utilization of non-stationary pulses with free parameters seems to be perspective for the series of short communication networks and optical cimputers. The major new features of this approach, as compared to the previous case, are the oscillating recovery of initial envelopes, the narrowing of solitary waves near by waveguide cutoff, the decelerated deformations of non-soliton pulses close to critical regimes, the coupled modulation of amplitude and apectral envelopes of optical signals, the power-dependent splitting of pulses. The rigorous general theory of such phenomena is abandoned in mathematics (Zakharov *et al.* 1980). To simplify the problem without the loss of generality some suitable analytical methods operating beyond of the scope of perturbation theory, such as variational approach, search of self-similar solutions and non-linear geometric optics, are described here. The gradually complicating concept of solitary pulses including the analysis of above-mentioned phenomena for sech, Gaussian and parabolic-shaped pulses is the main topic of Chapter 3.

The modal effects in thresholdless evolution of sech-shaped pulses are illustrated in Section 3.2. These effects are responsible for arising of both hierarchy of expanding TE_{0p} solitons and for peculiar TM_{01} soliton, narrowing near the cutoff. Unlike this the cross-modulation of polarization components of sech-shaped pulse travelling in T_{11} mode is shown to provide the self-compression and splitting of its amplitude-phase envelopes.

The power threshold phenomena in evolution of both soliton and non-soliton pulses is analysed in Section 3.3. These phenomena result in properties of high-order soliton with oscillating envelopes and formation of critical regime of propagation of partially stabilized non-soliton pulses. The perspectives of coding of combined amplitude and phase modulated signals using these pulses are marked. Unlike the theoretical model of infinitely extended solitons the influence of low intensity steepened and smoothened edges of real pulses with final duration on their dynamics is emphasized.

The expanding concept of oscillating solitary pulses due to initial controlled or random modulation is performed in Section 3.4. The tunable periodical recovery as well as the monotonous spreading of modulated pulses are shown, and the threshold values of parameters, separating these regimes, are revealed.

The non-linear shaping of pulses in the waveguide may be impeded due to generation of new optical harmonics. These processes are amplified sometimes due to waveguide dispersion. The dispersion-dependent conversion of pair or trio of guided waves connected with doubling and tripling of pumping frequency is discussed in Section 3.5. The interaction of pumping and the first-Stokes mode travelling with different velocities in a Raman waveguide provides the formation of a non-linear spatial scale characterizing the region of these modes coupling.

The governed self-shaping of pulses has a simple intuitive meaning, being considered in the framework of non-linear geometric optics for guiding systems (Section 3.6). The hydrodynamical analogy ensures the visuality of non-linear dynamics of localized wave distributions. The wide classes of exact analytical solutions obtained by this approach describe the formation of spike-like pulses as well as pulses stabilization by means of phase predistortions.

Unlike the self-action of solitary pulse investigated above, the interaction of a pair of pulses is discussed in Section 3.7. The modal effects in non-reciprocal phase shifts of colliding pulses are important for their fast phase modulation. The non-linear regimes, both impeding the overlapping of edges of adjacent co-propagating pulses due to their spreading and, vice versa, stimulating their self-trapping in optical circuits are illustrated.

The tumable narrowing, flattening and asymmetrical steepening of pulses useful for signals processing are considered in Section 3.8. The peculiar narrowing of the pulse, its maximum being invariant, and pulse maximum flattening, its edges being perturbed slightly, may be achieved by the use of non-linear polarization effects. The asymmetrical self-confinement of short pulses due to non-stationary process is described.

3.2 Non-linear Schrödinger equation for the lossless model of a waveguide

The non-linear evolution of wave pulse in a dispersive lossless medium may be considered as the result of stimultaneous development of two processes: phase self-modulation and dispersive deformation. The competition of these processes can reduce the rate of deformation and stabilize the primary pulses envelope. The dynamics of self-localized fields based on such competition and described by the master equation (3.11) with $Q = 0$ picks out an important type of solitary pulses, which maintain their shapes in the course of propagation. Undoubtedly, the most popular representation of such pulses is the soliton often considered as a standard model for the numerous self-localized distributions of intense fields of different physical nature.

Such amplitude–phase distribution in the simplest case related to the model of homogeneous linearly polarized wave in an unbounded medium may be written as

$$f = \frac{e^{i\eta\Delta_0^{-2}}}{\mathrm{ch}\left(\dfrac{q}{\Delta_0}\right)} . \tag{3.30}$$

Here Δ_0 is an effective width of the soliton, its peak amplitude being equal to unity:

$$\Delta_0 = \sqrt{\frac{2}{\chi}}. \tag{3.31}$$

The soliton's phase grows with distance η, while the amplitude envelope does not depend upon distance.

The real polarization structure of the waveguide modes containing two or three components of polarization can provide peculiar 'crosstalk' between these components, as was shown in Section 1.4. Taking this interaction into account, we can obtain a coupled system of non-linear Schrödinger equations for each of mode polarization components. Amplitude–phase cross-modulation of components impedes soliton formation in these field structures.

However, the problem of soliton existence is simplified for some modes. Thus TE_{0p} solitons containing only one polarization component are described by an equation similar to (3.11). Although the structure of the TM_{01} mode includes two polarization components, the phases of both components were shown to remain equal during non-linear evolution (see Section 1.4). This circumstance permits us to discuss TM_{01} solitons with the values of cutoff-dependent parameter

$y^2 = k_\perp^2 \beta^{-2}$ (1.141) being arbitrary by analogy with the family of TE_{0p} solitons. The set of non-linear Schrödinger equations governing the localized pulse in the TE_{11} mode shows the trend of amplitude–phase evolution in higher TE modes for the simplest hyperbolic-secant pulse (3.30). These trends are shown to provide the considerable pulse distortions in the vicinity of waveguide cutoff (Section 3.2.2).

3.2.1 *Hierarchy of solitons in multi-mode fibre far from cutoff*

An important peculiarity of non-linear wave processes in the multi-mode waveguide is connected with the existence of modal dispersion and mode-dependent polarization coefficients B_{np} given, for example, for the ring-shaped parabolic-index model of waveguide, in (3.15). The variety of non-linear susceptibilities of the guided system for a given non-linear susceptibility of the waveguide's material is determined by the different structure of the wave field in each mode. Such variety leads to the existence of the hierarchy of TE_{0p} solitons in a multi-mode waveguide. The effective widths of these solitons are

$$\Delta_{0p} = \frac{\Delta_0}{\sqrt{B_{0p}}} \tag{3.32}$$

with coefficients B_{0p} given in (3.15). The family of TE_{0p} solitons are depicted again by formula (3.30) after the substitution of Δ_{0p} (3.32) instead of Δ_0. This hierarchy, shown in Fig. (3.4), illustrates the broadening of solitons in higher TE_{0p} modes as compared with the fundamental TE_{01} mode, although the peak amplitudes of these pulses remain equal (Fig. 3.4). The dynamics of localized pulses travelling in a slab waveguide in the TE_{01} mode is described by the same master equation (3.8) with non-linear length L_n given in (1.11) and related values of the parameter $\chi = \chi_{01}$ and effective width $\Delta = \Delta_{01}$. The set of non-linear Schrödinger equations governing the evolution of a localized wave pulse travelling in a slab waveguide in the TM_{01} mode may be obtained by time-dependent generalization of the related stationary system (1.137):

$$i\frac{\partial E_i}{\partial \eta} + \frac{\partial^2 E_i}{\partial q^2} + \chi\left[\nabla_i\left(\frac{E_j\nabla_j|\mathbf{E}|^2}{\mathscr{E}}\right) + \frac{\omega^2}{c^2}(A|\mathbf{E}|^2E_i + BE^2E_i^*)\right] = 0. \tag{3.33}$$

The self-localized solution preserving the pulse envelopes for both E_x and E_z components of the TM_{01} mode (Fig. 1.16) may be obtain in a traditional soliton form (3.30) with an effective with depending upon the parameter $y^2 = k_\perp^2 \beta^{-2}$:

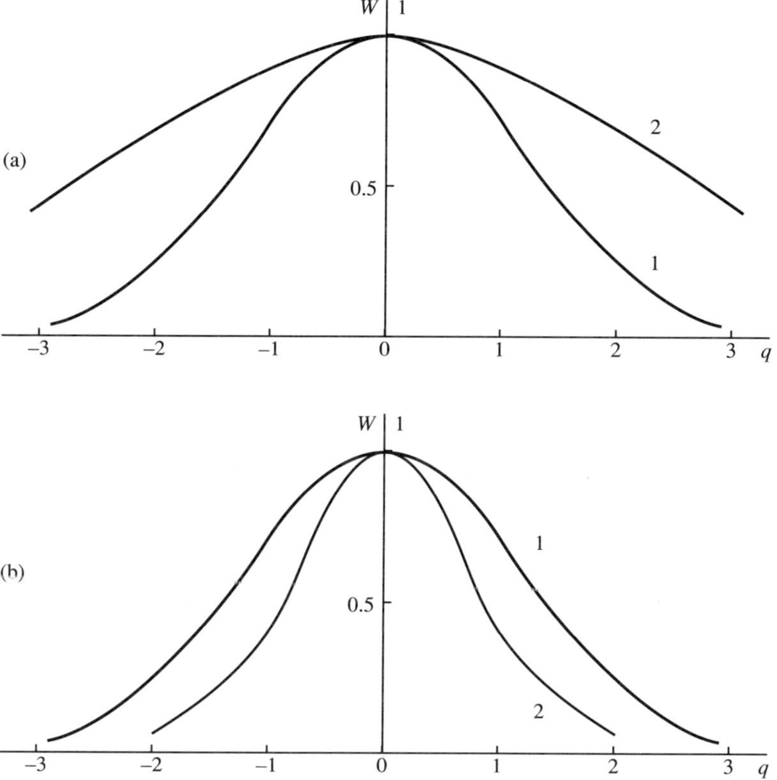

Fig. 3.4 (a) The discrete multitude of TE_{0p} solitons in a parabolically-shaped fibre (1.62). Curves 1 and 2 relate to TE_{01} and TE_{02} solitons $W(q)$, the peak amplitudes being equal. (b) The continuous dependence of TM_{01} solitons in a slab waveguide upon the parameter y. Envelopes 1 and 2 relate to the values $y = 0$ and $y = 1$, respectively.

$$\Delta(y^2) = \frac{\Delta_{01}}{\sqrt{F(y^2)}}. \tag{3.34}$$

The function $F(y^2)$ was calculated in (1.146):

$$F(y^2) = 1 + \frac{2y^2}{3} + y^4. \tag{3.35}$$

One can see that in the vicinity of cutoff ($y^2 \gg 1$) the effective width of the TM_{01} soliton narrows and its non-linear phase shift increases (Fig. 3.4). Far from cutoff ($y \to 0$) the longitudinal component of the electric field in the TM_{01} mode tends to zero, and the effective widths of TM_{01} solitons coincide, $\Delta(0) = \Delta_{01}$. It is useful to introduce the soliton's half-width $q_{0.5}$ calculated from the condition $|f(q_{0.5})| = 0.5$, that is

$$q_{0.5} = 1.32\Delta. \tag{3.36}$$

Thus the soliton envelope is characterized by a single parameter χ (3.1) for the given mode.

The factor of importance for use of solitons in fibre communication system is their spectral bandwidth. Using the Fourier transformation of envelope (3.30) at the fibre's input $z = 0$ (Hasegava *et al.* 1981),

$$\frac{1}{2T_0} \int_{-\infty}^{\infty} \frac{e^{-i\Omega t}dt}{\mathrm{ch}(t/T_0\Delta)} = \frac{\pi\Delta}{2} \frac{1}{\mathrm{ch}(\pi\Delta T_0\Omega/2)}, \tag{3.37}$$

we may find the spectral range $(-\Omega, \Omega)$ contains, for example, 99% of soliton energy:

$$\Omega T_0 = 1.6\Delta^{-1}. \tag{3.38}$$

This range usually defined as the spectral bandwidth, proves to be independent of the pulse duration $2T_0$. Substituting the value of Δ (3.32) into (3.38) we obtain

$$\Omega_{0p} = 2.9v_g\sqrt{\left(\left|\frac{\chi|E_0|^2}{v_\omega}\right|\frac{B_{0p}}{\lambda_0 n_0}\right)}. \tag{3.39}$$

The numerical coefficient arises here as the product of coefficients 1.6 from (3.38) and $\sqrt{\pi}$ included in the parameter χ (3.11).

Thus both spatial and spectral widths of the soliton are determined completely by its peak amplitude E_0 for given fibre characteristics and mode of propagation. This property will be shown further to determine the merits and shortcomings of use of solitons in fibre information systems.

3.2.2 *Mode polarization effects in self-compression of pulses close to cutoff*

The interaction of the polarization components of continuous radiation propagating in TE_{np} or TM_{np} waveguide modes was shown to produce unequal phase shifts of each component (Section 1.4). For the dynamics of the pulse travelling in the same modes, one can expect that this effect will provide the formation of different spectral modulation of each polarization component of the pulse. Such mode-dependent chirp, the dispersion being taken into account, can give rise to complicated self-consistent tuning of the polarization structure of the pulse itself and the amplitude and frequency envelopes of its components.

Let us discuss these phenomena for pulses travelling in a rectangular waveguide in the TE_{11} mode. Starting with the generalized vector equation (3.33) and making use of the analytical approach developed in Section 1.4, we obtain the system of coupled equations governing the behaviour of functions f_1 and f_2 related to the E_x and E_y polarization components, respectively:

$$i\frac{\partial f_1}{\partial \eta} + \frac{\partial^2 f_1}{\partial q^2} + \chi f_1\{3|f_1|^2 y_2^2[3(1+y^2) - 2y^2] + |f_2|^2 y_1^2(1 + y^2 + 2y_1^2)\}$$

$$+ 2\chi f_2 y_1^2(3|f_2|^2 y_1^2 - |f_1|^2 y_2^2) = 0; \tag{3.40}$$

$$i\frac{\partial f_2}{\partial \eta} + \frac{\partial^2 f_2}{\partial q^2} + \chi f_2[\,|f_1|^2 y_2^2(1 + y^2 - 2y_2^2) + 3|f_2|^2 y_1^2[3(1+y^2) + 2y_2^2]\,]$$

$$- 2\chi f_1 y_2^2(3|f_1|^2 y_2^2 - |f_2|^2 y_1^2 = 0. \tag{3.41}$$

Parameter χ in this system is determined as before (3.13) by the ratio of dispersive length L_ω and non-linear length L_n. To emphasize the cutoff effect, the dispersive length L_ω for the TE mode may be expressed via the waveguide parameter $y^2 = k_\perp^2 \beta^{-2}$.

$$(L_\omega)_{TE} = \frac{\sqrt{1 + y^2}}{\beta_0}\left[1 - \frac{2\omega^2 \mathscr{E}}{1 + y^2} \frac{\dfrac{\partial^2(\omega^2 \mathscr{E})}{\partial \omega^2}}{\left[\dfrac{\partial(\omega^2 \mathscr{E})}{\partial \omega}\right]^2}\right]. \tag{3.42}$$

Here β_0 is the wavenumber for an unbounded dielectric medium ($\beta_0 = 2\pi\lambda_0^{-1}n_0$); the non-linear length L_n for the TE_{11} mode is given by (1.151).

Some results for the temporal evolution of a hyperbolic-secant pulse launched to the waveguide's input,

$$f_1|_{\eta=0} = f_2|_{\eta=0} = \frac{1}{chq}, \tag{3.43}$$

$$\varphi_1|_{\eta=0} = 0; \qquad \varphi_2|_{\eta=0} = 0, \tag{3.44}$$

are illustrated in Fig. 3.5. To compare these non-stationary effects with the related stationary ones discussed in Section 1.4, the same values of geometric parameters $y_2^2 = 0.09, y_1^2 = 0.36$ are used (see Fig. 1.19). One can see that amplitude–phase evolution of the TE_{11} pulse shown in Fig. 3.5 is possessed of oscillatory properties.

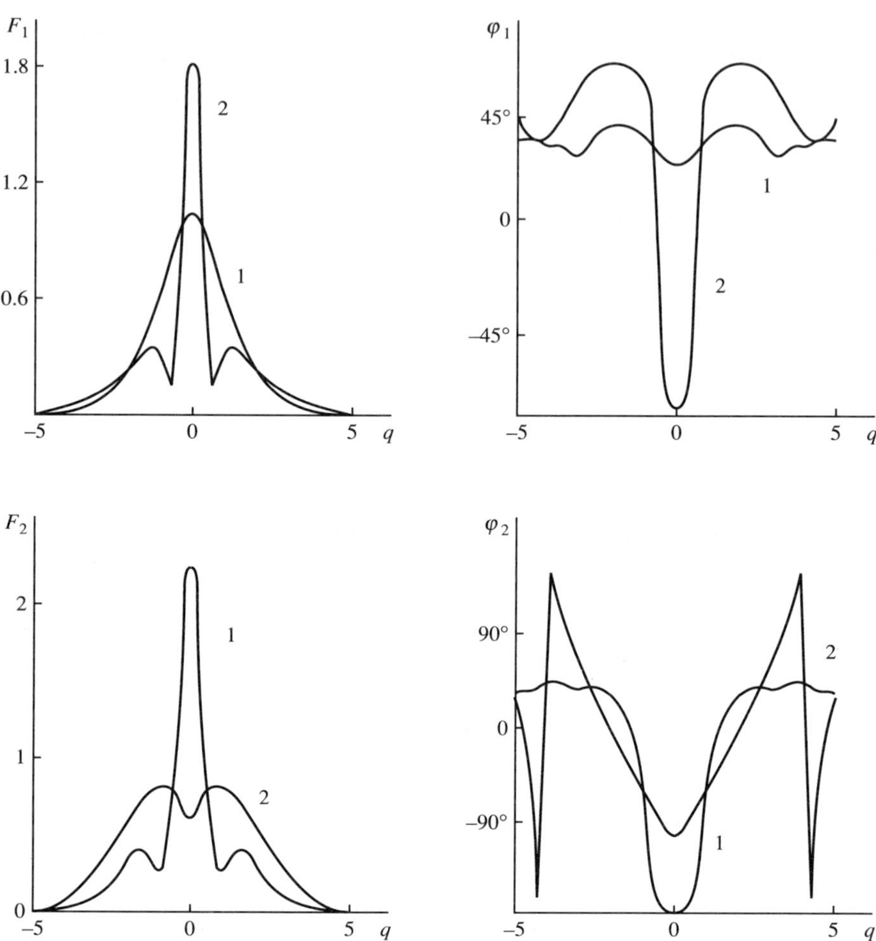

Fig. 3.5 Cross-modulation of polarization components and phases of a hyperbolic-secant pulse (3.43, 3.44) in the TE_{11} mode in the rectangular waveguide shown in Fig. 1.12. Curves 1 and 2 illustrate the amplitude $F_{1,2}$ and phase $\varphi_{1,2}$ envelopes at positions $\eta = 0$ and $\eta = 0.8$, respectively.

3.3 Threshold phenomena in non-linear pulse dynamics

The development of the physics and techniques of short light pulses stimulates the elaboration of different modifications of all-optical high-bit-rate long-distance communication and information systems. Success in this field depends essentially upon the progress in formation of powerful wave pulses and their stabilization in the course of propagation in optical channels. To use some non-linear properties of such pulses discussed above, it is necessary to choose two important characteristics of such a channel:

1. the system of coding of information;
2. the elementary symbol for this system.

Some phenomena determining the information rate I of a non-linear information channel in the simplest noiseless model may be evaluated by means of the well-known Shannon formula written in the form

$$I = R \, \log_2 M. \tag{3.45}$$

Here R is the number of signals transmitted per second; the factor M is the number of states of these signals. The minimum value of factor M, relating to either existence or absence of the signal in the given time interval, is equal to $M = 2$. Thus, the utilization of soliton signals is based on the value $M = 2$, because the soliton pulse (3.30) has no additional degrees of freedom owing to the connection between its half-width and peak amplitude. Attempts at non-linear stabilization of elementary signals in waveguide channels have produced a variety of proposals for the types of signals and codes utilized. These may influence on both R and M factors in (3.45).

The tendency to growth of the information rate I (3.45) proportionally to factor R, via narrowing of the pulses and shortening of the distances between adjacent ones, is restricted by pulse diffusion and interaction. On the other hand, the weak logarithmic dependence of I (3.45) upon the M factor cannot change appreciably the information rate in quantitative terms. However, some qualitative peculiarities of this information channel provided by the coding system and elementary symbols are connected via the factor M.

The familiar transmission of messages in optical fibres by means of soliton signals using the double-valued code ($M = 2$) may be characterized by values of R factor determined by the pulses half-width, the distance of propagation, and the pulse interaction. Some properties of these elementary symbols were described above in Section 3.2.1. The logarithmic factor in eqn (3.45) may become important for non-soliton pulses ($M > 2$). Unlike solitons, the energy of these pulses does not depend upon their half-widths. This energy, given, for example, by the the shape of pulse, with fixed peak amplitude and duration, may be used for identification of non-soliton signals in the receiver, as long as the spreading of an information-carrying pulse in the fibre can be ignored. The combination of energy variations of the signals with the displacement modulation shows the particular possibilities of such combined coding:

1. The creation of datamation systems based on many-valued logic. Thus, the simplest case of triple-valued logic may be realized by means of two non-soliton signals; the more powerful signal corresponds to $+1$, the less powerful one to -1; the absence of any signal indicates zero.

2. The transmission of probabilistic information. The variable energy of the pulse, with constant distance between pulses, shows the possibility of coding the probability of the event connected with the fixed signal.

3. The transmission of complex numbers. This possibility, based, for example, on the simultaneous and independent coding of amplitude and phase of the complex number by means of displacement modulation of the signals and their energy variation, seems to be especially attractive.

These perspectives are interesting for the comparatively short waveguide channels ($z \leqslant L_\gamma$) utilized, e.g., for communication between computers inside the 'fibred city'. The penalty one must pay for the growth of information rate with the utilization of non-soliton pulses as the elementary symbols is imposed the limitation of the distance of distortion-free propagation of such signals. On the other hand, soliton pulses are often considered as the elementary symbols for distant communication ($z \gg L_\gamma$): however, the information-carrying properties of such channels are limited due to utilization of the double code that is typical for soliton-like signals. Thus, unlike linear information systems, the choices of elementary symbols and of the code for transmission in non-linear information channels prove to be dependent upon each other.

The energy W of fundamental solitons does not depend upon any energetic threshold,

$$W = \int\limits_{-\infty}^{\infty} \frac{\mathrm{d}q}{\mathrm{ch}^2(q/\Delta)}. \tag{3.46}$$

This energy W could be coupled into the waveguide by means of a series of pulses with different durations $2T_0$ and peak amplitudes E_0 satisfying the condition

$$T_0|E_0|^2 = \text{const.} \tag{3.47}$$

All these fundamental solitons remain invariant in the course of propagation in the loseless medium. The influence of energy threshold on the evolution of high-order soliton and non-soliton envelopes is discussed further.

3.3.1 *High-order solitons in single-mode fibres*

In any realistic system it may not be possible to control the power launched into an optical fibre to ensure that an initial pulse shape of the fundamental soliton form (3.30) is obtained. Thus it is instructive to analyse the dynamics of pulses,

$$f_N(q)\big|_{\eta=0} = \frac{N}{\mathrm{ch}(q/\Delta)}.$$
(3.48)

The positive integer values of parameter $N (N = 2, 3, 4, \ldots)$ relate to so-called high-order solitons. In the case $N = 2$, the exact analytical solution of non-linear Schrödinger equation obeying the condition (3.48) may be written (Doran and Blow 1983) as

$$f_2(q,\eta) = \frac{4e^{-i\eta\Delta^{-2}/2}[\mathrm{ch}(3q/\Delta) + 3e^{-2i\eta\Delta^{-2}}\mathrm{ch}(q/\Delta)]}{\mathrm{ch}(4q/D) + 4\mathrm{ch}(2q/\Delta) + 3\cos(2\eta/\Delta^2)}.$$
(3.49)

Unlike the fundamental soliton ($N = 1$), which is independent of η, the second-order soliton (3.49) develops its structure periodically in η (Fig. 3.6). The evolution of the pulse $f_2(q,\eta)$ leads to self-lamination of the localized field at the point (q_0, η_0),

$$q_0 = 0.66\Delta, \qquad \eta_0 = \frac{\pi\Delta^2}{2}.$$
(3.50)

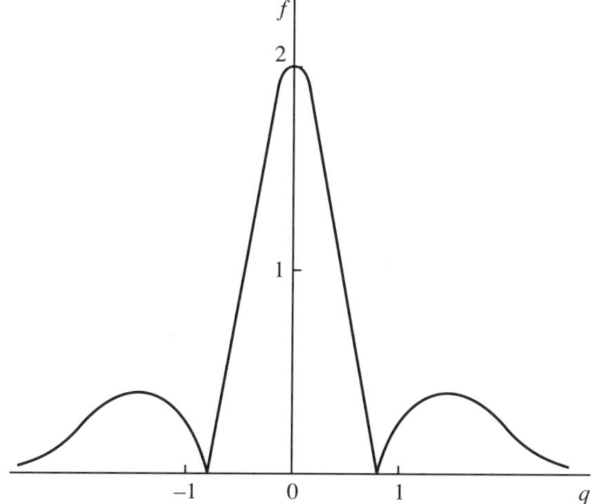

Fig. 3.6 Periodic self-stratification of an $N = 2$ soliton.

The pulse shape at $\eta = \eta_0$ has three peaks, which lie within the original pulse width. The initial envelope is reinstated at the point η_0, which defines the spatial period of oscillation:

$$\eta_0 = \pi \Delta^2. \tag{3.51}$$

We can also generate a continuous set of solitons from f_1 (3.30) and f_2 (3.49) by using the invariance of the non-linear Schrödinger equation under the following transformation:

$$\eta = Q^2 \eta; \qquad q = Qq; \qquad f = \frac{f}{Q}. \tag{3.52}$$

The derivation of analytical formulae for high-order solitons with $N > 2$ is difficult. The results (3.49–3.52) are valid for pulses travelling in both fundamental and higher modes of the TE_{0p} waveguide, the related value of parameter Δ_{0p} being taken into account.

It is important to point out that the use of higher-order solitons leads to their periodical stratification; thus distortion-free information may be received in this system only at some discrete points $\eta = \eta_0$ along the waveguide. In contrast, the use of non-soliton pulses is shown below to provide the possibility of receiving stabilized signals at arbitrary points inside the restricted span of the waveguide.

3.3.2 *Subcritical and supercritical regimes of propagation of non-soliton pulses*

Some important trends of temporal evolution of powerful pulses may be revealed by means of general properties of eqn (3.11). Noting the formal equivalence of this equation with the system of equations describing scattering on the potential $U = |f|^2$,

$$\frac{\partial f_1}{\partial q} = \mathrm{i} \sqrt{\frac{\chi}{2}}\, U f_2, \tag{3.53}$$

$$\frac{\partial f_2}{\partial q} = \mathrm{i} \sqrt{\frac{\chi}{2}}\, U f_1 + \mathrm{i} \lambda f_2, \tag{3.54}$$

and considering λ to be the spectral parameter, one can use so-called 'inverse scattering method' (Zakharov *et al.* 1980). Basically, direct use of this method demands a high level of mathematical analysis, while some quantitative conclusions may be derived straightforwardly from the system (3.53–3.54).

Let us introduce a new function $V(q)$ connected with the amplitude $f_0(q)$ of the unmodulated pulse at the waveguide input $\eta = 0$:

$$V(q) = \int_{-\infty}^{q} f_0(x)\,\mathrm{d}x. \tag{3.55}$$

Supposing the functions f_1 and f_2 to be defined at $q = -\infty$ as

$$f_1|_{q=-\infty} = 1, \qquad f_2|_{q=-\infty} = 0, \tag{3.56}$$

we may investigate the asymptotic behaviour of these functions at $q \to \infty$, reducing the system (3.53, 3.54) with $\lambda = 0$ to one equation:

$$\frac{\partial^2 f_1}{\partial V^2} + \frac{\chi}{2} f_1 = 0. \tag{3.57}$$

The functions f_1 and f_2 are connected via the relation

$$|f_1|^2 + |f_2|^2 = 1. \tag{3.58}$$

The solution of eqn (3.57) satisfying condition (3.56) is

$$f_1 = \cos\left[\sqrt{\chi/2}\, V(q)\right]. \tag{3.59}$$

The amplitude of the first fundamental scattering maximum described by function f_1 (3.59) relates to the maximum difference between the initial state ($q = -\infty, f_1 \to 1$) and the asymptotic state ($q \to \infty$, $f_1 \to 0$). The threshold value of the parameter $\chi = \chi_{\mathrm{cr}}$ providing the formation of this asymptotic state follows from eqn (3.59):

$$\chi_{\mathrm{cr}} = \frac{\pi^2}{2} \left| \int_{-\infty}^{\infty} f_0(q)\,\mathrm{d}q \right|^{-2}. \tag{3.60}$$

This result defines the asymptotic behaviour of the pulse via its initial amplitude profile. The threshold value $\chi = \chi_{\mathrm{cr}}$ separates two types of evolution of pulses with similar shapes $f_0(q)$ with respect to their peak amplitudes E_0. The evolution of subcritical pulses ($\chi < \chi_{\mathrm{cr}}$) is characterized by deceleration of dispersive spreading due to the influence of non-linearity, although the power of the pulse is unsufficient to stabilize the spreading completely. In contrast, the supercritical regime ($\chi > \chi_{\mathrm{cr}}$) provides self-compression of the localized field. The critical value $\chi = \chi_{\mathrm{cr}}$ relates to the asymptotic formation of a stabilized field distribution.

However, although revealing the tendency of evolution of the input pulse $f_0(q)$, eqn (3.60) contains no information about the spatial scales of this evolution. Thus, to obtain the complete picture of deformation of the input pulse in a stabilized regime we must solve the master equation (3.11) for the value $\chi = \chi_{\mathrm{cr}}$ and initial envelope $f_0(q)$. Some examples of such a procedure are illustrated here. The application of formula (3.60) to the one-parameter family of bell-like pulses characterized, unlike solitons, by the final duration $(-T_0; T_0)$,

$$f_0(q) = \begin{cases} (1 - q^2)^{\nu/2}, & q^2 \leqslant 1, \\ 0, & q^2 \geqslant 1, \end{cases} \tag{3.61}$$

leads to a simple result:

$$\chi_{\text{cr}} = \frac{\pi}{2} \frac{\Gamma^2[(\nu + 3)/2]}{\Gamma^2[(\nu + 2)/2]}. \tag{3.62}$$

Here Γ is the gamma-function. The pulses (3.61) are shown in Fig. 3.7; the parameter ν determines the pulse's half-width:

$$q_{0.5} = \sqrt{1 - 2^{-2/\nu}} \tag{3.63}$$

The result (3.62) is valid for an arbitrary positive value of the parameter $\nu \geqslant 0$ (Fig. 3.8). The minimum value $\chi_{\text{cr}} = \pi^2/8$ relating to the rectangular pulse ($\nu = 0$) was obtained formerly by Zakharov *et al.* (1980). The variations of dimensionless energy of these pulses, W_ν,

$$W_\nu = \int\limits_{\infty}^{\infty} |f_0(q)|^2 \, \mathrm{d}q \tag{3.64}$$

are shown in Fig. 3.9.

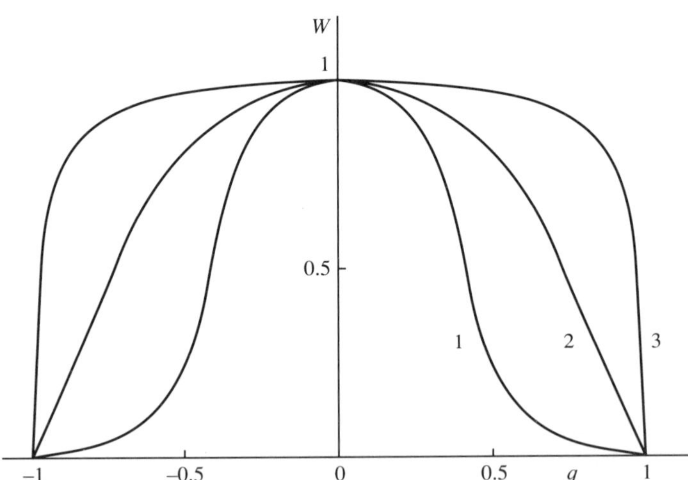

Fig. 3.7 The intensity envelopes of non-soliton pulses (3.61) with coinciding maxima $W_{\text{max}} = 1$ confined in the interval $-1 \leqslant q \leqslant 1$ distinguished due to values of the free parameter $\nu = 2$ (curve 1), $\nu = 0.5$ (curve 2), $\nu = 0.125$ (curve 3). The decrease of parameter ν corresponds to steepening of pulse fronts.

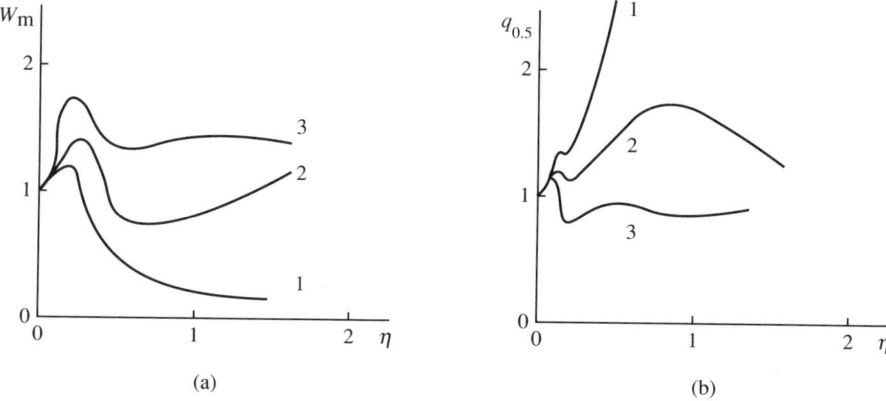

Fig. 3.8   The subcritical and critical regimes of pulse (3.61) evolution ($\nu = 2$). (a) and (b) show the dependence of the pulse peak intensity W and half-width $q_{0.5}$ upon the propagation distance η. Curves 1, 2, and 3 relate to the linear ($\chi = 0$), subcritical ($\chi = 0.5\chi_{cr}$) and critical ($\chi = \chi_{cr}$) propagation regimes, respectively.

The propagation of the pulse (3.61) with $\nu = 2$, obtained by numerical solution of eqn (3.11), is illustrated in Fig. 3.9 for critical and subcritical regimes. One may see that the self-localized distribution is formed in the vicinity of the pulse maximum ($\eta \leqslant 0.3$). The subsequent propagation in such a critical regime is not accompanied by any considerable deformation of the pulse at distances $\eta \leqslant 2$. In the subcritical regime of the same pulse propagation ($\chi < \chi_{cr}$, the other parameters being equal), dispersive diffusion will develop, although the pace of such diffusion will be slackened in comparison with the 'linear' diffusion. The supercritical regime of pulse evolution is

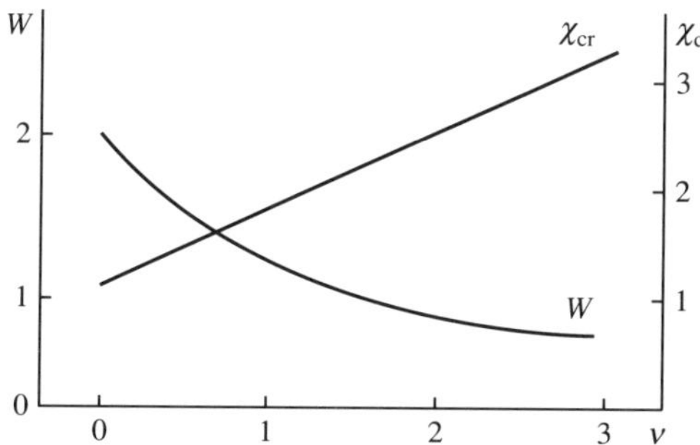

Fig. 3.9   Threshold parameter χ_{cr} (3.62) and dimensionless energy $W\nu$ (3.64) of non-soliton pulses (3.61) are plotted against the value of the free parameter ν.

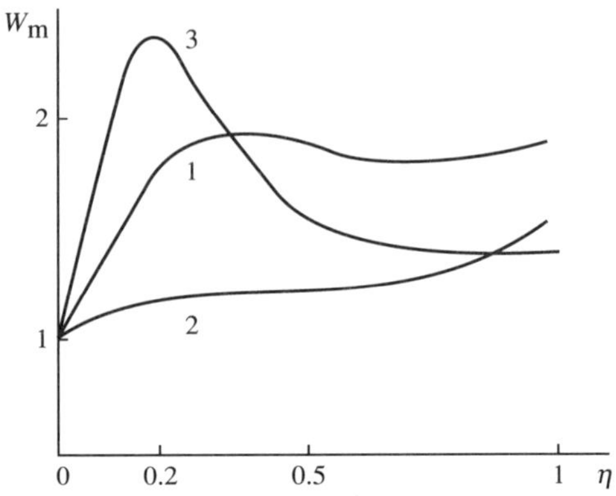

Fig. 3.10   The influence of initial modulation on evolution of a parabolically shaped pulse (3.65) with $\nu = 2$. The pulse peak intensity W_m is plotted against the distance of propagation η. Curves 1, 2, and 3 relate to the values of modulation parameter $\delta = 0$, -1, and 1, respectively.

characterized by its maximum increase and formation of new maxima due to stratification of the envelope.

The dynamics of formation of self-localized distributions may be varied by the frequency 'chirp' of the initial pulses. The initial 'chirp', depending on its distribution along the pulse, may weaken or strengthen the 'chirp' arising in the course of pulse propagation due to its self-modulation. This change of frequency modulation has an influence on the rate of amplitude rebuilding. Some tendencies of these phenomena are illustrated in Fig. 3.10 for modulated pulses,

$$f_0(q) = \begin{cases} (1 - q^2)^{1/2} \exp(i\delta q^2), & q^2 \leqslant 1, \\ 0, & q^2 \geqslant 1. \end{cases} \tag{3.65}$$

The evolution of pulses (3.61) with given peak amplitude and duration, parameter ν being free, may be weakened by condition $\nu \leqslant 4$ related to the 'smoothed' edges of pulses (Fig. 3.11). These smoothed pulses, unlike the stationary ones, have no discontinuities of derivaties of the field at the edges; such a shape attenuates the diffusion of these pulses.

Thus the peak amplitude and half-width of non-soliton bell-like pulses (3.61) with the final duration can be choosen in order to stabilize these pulses at distances of propagation of about (1–2) L_ω. Any amplitude sufficient for the critical regime of pulse propagation in the given mode may become insufficient for the analogous regime in another mode, for equal pulse durations. It is important that the

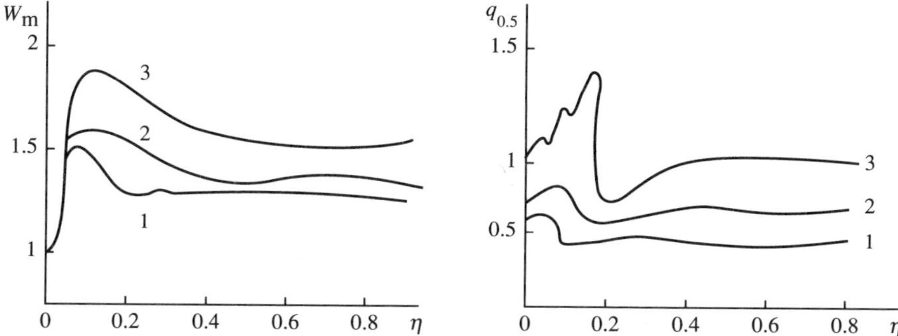

Fig. 3.11   The slowing of the deformation of the pulse (3.61) in the critical regime ($\chi = \chi_{cr}$) due to smoothing of the envelope's edges determined by parameter ν. (a) and (b) illustrate the dependence of pulse intensity maximum W_{m} and its half-width $q_{0.5}$ upon the distance of propagation η. Curves 1, 2, and 3 relate to the values $\nu = 5$, 3, and 1, respectively.

stabilized pulses with different energies, e.g. $W = 1$ ($\nu = 0.125$) and $W = 0.5$ ($\nu = 4$), may be used as elementary symbols in all-optical information systems for coding of complicated blocks of information.

3.4   Controlled and random modulation in the stabilization of localized wave fields

The trends and rates of non-linear evolution of localized wave pulses in a waveguide depend essentially upon the amplitude-phase envelopes of this pulse at the waveguide's input. Such sensitivity attracts attention to the accuracy of shaping of these envelops. Along with controlled formation of amplitude–phase structure of the input pulse, the random perturbations of these distributions produced by real lasers and modulators have an influence on the coupled non-linear evolution of this structure in fibres. Since such a process is often governed by primary phase modulation it is useful to analyse the influence of both controlled (Section 3.4.1) and random (Section 3.4.2) phase distributions on the dynamics of localized wave fields. The flexibility of phase tuning can allow a variety of tendencies and scales of pulse deformation. Some phase distributions are shown to produce spatial oscillations of travelling pulses and periodical recovery of their initial profiles and peak amplitudes. The use of such pulses in optical information channels determines the spacing of receivers in this system (Mollenauer *et al.* 1991).

To obtain the analytic self-consistent description of these coupled processes, the approximate methods of solution of the non-linear Schrödinger equation (3.11) are demonstrated below. These methods are based on the use of self-similar solutions. Computer simulation shows that the tendencies of evolution given by self-similar solutions are valid as long as the distances of propagation are restricted to several dispersion lengths L_ω. Another method of approximate analytic solution of the non-linear Schrödinger equation free of these limitations is connected with non-linear geometric optics (see Section 3.6).

3.4.1 *Non-linear oscillations of modulated solitons*

Unlike phase-unmodulated fundamental solitons that conserve their amplitude envelope (3.30), initial phase modulation can provide self-similar evolution of the soliton pulse. Analogously to modulated non-soliton pulses (3.65), the same initial modulation described by the exponential factor $\exp(i\delta q^2)$ is expected to accelerate or impede the soliton evolution subject to the sign of parameter δ. However, unlike the envelopes (3.65), the soliton pulses exhibit the possibility of periodically reinstating their initial (hyperbolic-secant)^2 shape. To analyse the dynamics of these non-linear coupled amplitude–phase oscillations it is instructive to use the variational approach (Anderson 1983).

Such an approach is based on utilization of the special function $L(\mathbf{E};$ $\mathbf{E}^*;$ $\nabla\mathbf{E};$ $\nabla\mathbf{E}^*)$ characterizing the electromagnetic field $\mathbf{E}$ and the extremal properties of the integral

$$I = \int_0^q L\, dq_1. \tag{3.66}$$

The well-known condition determining the spatial evolution of the field distribution described by these integrals I may be written as an Eulerian equation:

$$\frac{\partial}{\partial q}\left[\frac{\partial L}{\partial\left(\frac{\partial E_j}{\partial q}\right)}\right] = \frac{\partial L}{\partial E_j}. \tag{3.67}$$

The Laplace function L must be choosen to provide the derivation of the non-linear Maxwell equation (3.11) from the condition (3.67). Such a Lagrange function for the pulse $\mathbf{E} = \mathbf{E}_0 f(q,\eta)$ in TE_{0p} mode may be written in the form

$$L = \int_{-\infty}^{\infty} dq\left\{\left|\frac{\partial f}{\partial q}\right|^2 + i\left(f\frac{\partial f^*}{\partial\eta} - f^*\frac{\partial f}{\partial\eta}\right) - \frac{\chi}{2}\,|f|^4\right\}. \tag{3.68}$$

To reveal the conditions of existence of modulated solitons we can consider the generalized model of a soliton-like envelope described by variable peak amplitude $A(\eta)$, half-width $\Delta(\eta)$ and phase factor $b(\eta)$:

$$f(\eta, q) = \frac{A(\eta)e^{iq^2 b(\eta) + i\theta(\eta)}}{\mathrm{ch}\left[\dfrac{q}{\Delta(\eta)}\right]}. \qquad (3.69)$$

In the above-mentioned case of unmodulated profile of a fundamental soliton these functions are reduced to well-known constant values

$$A = 1, \qquad b = 0, \qquad \Delta = \Delta_0 = \sqrt{2/\chi}, \qquad (3.70)$$

and the function

$$\theta(\eta) = -\Delta_0^{-2}\eta \qquad (3.71)$$

The spatial evolution of these parameters may be analysed by substitution of the trial function (3.69) into the variational equations (3.67, 3.68). Introducing the normalized half-width of the pulse by means of function $y(\eta)$

$$\Delta(\eta) = \Delta_0 y(\eta) \qquad (3.72)$$

satisfying the condition

$$y(\eta)\big|_{\eta=0} = 1, \qquad (3.73)$$

we can obtain these equations in the form

$$yA^2 = 1; \qquad (3.74)$$

$$\frac{\partial\theta}{\partial\eta} = \frac{1}{\Delta_0^2}\left(\frac{1}{y^2} - \frac{5}{\pi y}\right); \qquad (3.75)$$

$$\frac{1}{y}\frac{\partial y}{\partial\eta} = -4b(\eta); \qquad (3.76)$$

$$\tfrac{1}{2}\left(\frac{\partial y}{\partial\eta}\right)^2 + \frac{\mu}{y^2} + \frac{\nu}{y} = 8b_0^2 + \mu + \nu. \qquad (3.77)$$

Here the value b_0 characterizes the initial modulation $(b_0 = b\big|_{\eta=0})$; parameters μ and ν are determined by the trial function

$$\mu = \frac{8}{\pi\Delta_0^4}; \qquad \nu = -2\mu. \qquad (3.78)$$

The variations of the pulse's half-width $y(\eta)$ are given by solution of eqn (3.77):

$$\eta = F(y),$$

$$F(y) = \frac{1}{4\sqrt{[(\pi\Delta_0^4)^{-1} - b_0^2]}} \int_1^y \frac{y\,dy}{\sqrt{[(y_1 - y)(y - y_2)]}}, \qquad (3.79)$$

$$y_{1,2} = (1 \pm \sqrt{\pi}\,|b_0|\Delta_0^2)^{-1}. \qquad (3.80)$$

Formula (3.79) describes the periodic narrowing and spreading of a modulated soliton-like pulse. Thus in the case $b_0 > 0$ one can conclude from (3.79) that the pulse narrows at the beginning of evolution from the value $y = 1$ to $y = y_1 < 1$ (Fig. 3.12). The subsequent increase of half-width up to the value $y = 1$ restores the initial envelope. The spatial half-period of this oscillation is

$$\eta_p = \tfrac{1}{2}F(y_1). \qquad (3.81)$$

In contrast, the initial modulation $b_0 < 0$ provides periodical broadening of the pulse from $y = 1$ up to $y_2 > 1$. Unlike the higher-order solitons (3.49), the modulated solitons do not split in the course of evolution.

It is important to point out that phase-modulated solitons do exist, the modulation being limited:

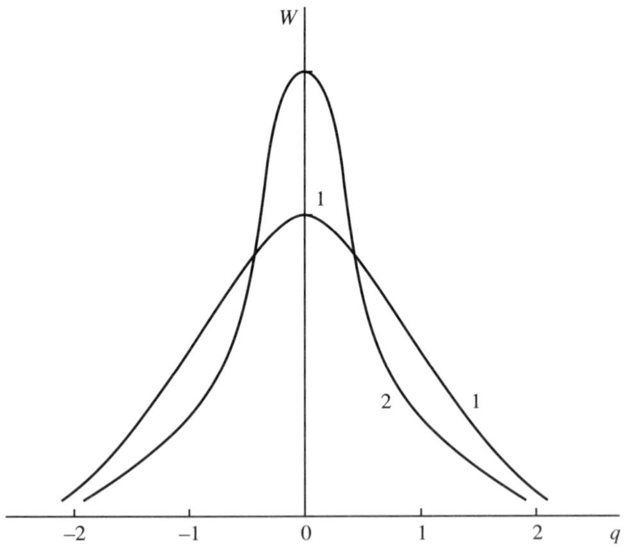

Fig. 3.12 Non-linear oscillations of a modulated soliton (3.69) with $b_0 = \sqrt{\pi}$, $\Delta_0^2 = 2$. Curves 1 and 2 relate to positions $\eta = 0$ and $\eta = \eta_p$ (3.81), respectively.

$$-\frac{1}{\sqrt{\pi\,\Delta_0^2}} < b_0 < \frac{1}{\sqrt{\pi\,\Delta_0^2}}. \qquad (3.82)$$

The case $b_0^2 \geqslant (\pi\Delta_0^4)^{-1}$ relates to unlimited spreading of the initial pulse.

The coupled amplitude–phase evolution results in periodic oscillations of the peak amplitude of modulated signals. Therefore, such signals may be interesting for all-optical systems using the intensity threshold discriminators for selection of information.

3.4.2 *Temporal coherency in the dynamics of short wave pulses*

The examples of coupled amplitude–phase self-modulation analysed above relate to the model of sufficiently long pulses: the pulse duration $2T_0$ was assumed to exceed the temporal coherence interval t_c. The duration of the picosecond pulses important for fibre communication may be comparable with the time scale t_c; therefore, the coherent properties of the input radiation may have a great influence on the dynamics of such short pulses.

The fluctuations of the optical radiation source and the errors of pulse shaping in the modulation system may be described by some random function $P(q)$. The complex envelope of the input signal may be written as the product

$$f(q) = f_0(q)P(q). \qquad (3.83)$$

Here the function $f_0(q)$ determines the initial pulse envelope, the fluctuations being ignored. Let us suppose that the correlation function $C(q)$ for the random process $P(q)$ is described by Gaussian statistics. This supposition is often justified for semiconductor lasers and photodiodes and leads to the condition

$$C(\theta) = \left\langle P\left(q - \frac{\theta}{2}\right)P^*\left(q + \frac{\theta}{2}\right)\right\rangle = B_0\exp\left(-\frac{\theta^2}{\tau_c^2}\right). \qquad (3.84)$$

The dimensionless parameter τ_c characterizes the influence of temporal coherency of the source on the evolution of the pulse with duration $2T_0$:

$$\tau_c^2 = \frac{t_c^2}{T_0^2}. \qquad (3.85)$$

The brackets $\langle\ \rangle$ denote an ensemble averaging process. The evolution of the pulse inside the waveguide may be described by means of the correlation function

$$C(\eta; q; \theta) = \left\langle f\left(\eta, q - \frac{\theta}{2}\right) f^*\left(\eta, q + \frac{\theta}{2}\right)\right\rangle \tag{3.86}$$

governed by eqn (3.87) obtained from the non-linear Schrödinger equation (3.11) for the lossless medium $(Q = 0)$:

$$i\frac{\partial}{\partial\eta} C(\eta, q, \theta) + 2\frac{\partial^2}{\partial q \partial\theta} C(\eta, q, \theta) + 2\chi\left[C\left(\eta, q + \frac{\theta}{2}\right)\right.$$

$$\left. - C\left(\eta, q - \frac{\theta}{2}\right)\right] C(\eta, q, \theta) = 0. \tag{3.87}$$

Considering the initial Gaussian pulse

$$f_0(q) = \exp\left(-\frac{q^2}{2}\right) \tag{3.88}$$

we can write the correlation function C at the input of the fibre in the form

$$C\big|_{\eta = 0} = C_0 \exp\left(-q^2 - \frac{\theta^2}{\tau_0^2}\right), \tag{3.89}$$

$$\tau_0^2 = \frac{4\tau_c^2}{4 + \tau_c^2}. \tag{3.90}$$

Some tendencies of non-linear evolution of the function $C(\eta, q, \theta)$ governed by eqn (3.87) may be illustrated by means of approximate self-similar solution of this equation satisfying the boundary condition (3.89):

$$C(\eta, q, \theta) = F(\eta) \exp\left[-\frac{q^2}{g(\eta)} - \frac{\theta^2}{\tau_0^2 h(\eta)} - i\alpha(\eta)q\theta\right];$$

$$\alpha(\eta) = -\frac{1}{2F}\frac{dF}{d\eta}; \qquad g(\eta) = h(\eta) = F^2(\eta). \tag{3.91}$$

Here the function $F(\eta)$ is obeys the equation

$$F^3\frac{d^2F}{d\eta^2} + 8\chi F - \frac{16}{\tau_c^2} = 0 \tag{3.92}$$

with the initial conditions

$$F\big|_{\eta = 0} = 1; \qquad \frac{dF}{d\eta}\bigg|_{\eta = 0} = 0. \tag{3.93}$$

The solution of eqn (3.92) satisfying condition (3.93) may be written as

$$\int\limits_{1}^{F} \frac{\mathrm{d}F}{\sqrt{\left(1 - \frac{1}{F}\right)\left(\frac{1}{F} - \frac{1}{F_0}\right)}} = \frac{4\tau\left|1 - \chi\tau_c^2\right|^{1/2}}{\tau_c}\,, \tag{3.94}$$

$$F_0 = (\chi\tau_c^2 - 1)^{-1}. \tag{3.95}$$

The pulse's behaviour proves to be dependent upon the ratio $\tau_c = t_c T_0^{-1}$. The small non-linearity $(\chi \leqslant \tau_c^{-2})$ relates to the monotonic spreading of the pulse, the larger χ values $(\chi > \tau_c^{-2})$ correspond to the pulse half-width F oscillations between the values $F = 1$ and $F = F_0$ (Fig. 3.13). The dimensionless spatial period of these oscillations η_p is equal to

$$\eta_p = \frac{\pi\chi\tau_c^3}{\left|\chi\tau_c^2 - 1\right|^{3/2}}. \tag{3.96}$$

Thus the final time scale of temporal coherency t_c of the input pulse determines the trend to pulse spreading in the course of propagation. The non linear self-compression is shown to impede this trend and to stabilize the pulse, providing its spatial oscillations. Owing to the decrease of the non-linear parameter χ for the higher waveguide modes (3.13), the peak amplitudes under the conditions of stabilization for these modes deteriorate.

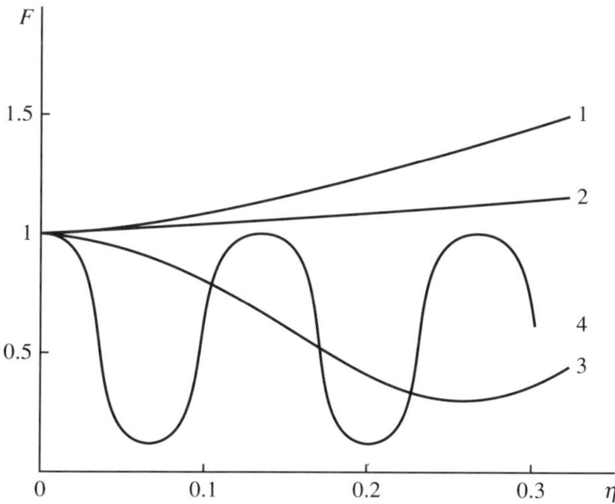

Fig. 3.13 Monotonic and oscillative regimes of power-dependent evolution of the correlation factor $F(y)$ (3.91) of the pulse through the temporal coherency parameter $\tau_c = 1.15$. Curves 1, 2, 3, and 4 relate to the values of the non-linear parameter $\chi = 1, 2, 5,$ and 50, respectively.

3.5 Dispersion-dependent absorption of powerful waves

The main contributions to the waveguide linear attenuation are produced by absorption and scattering. Owing to advances in fibre manufacturing techniques the Rayleigh scattering is connected with the dominant contribution to attenuation over most of the visible and near IR spectral range apart from OH absorption peaks, which are primarily at $\lambda_0 = 1.38\,\mu$m. For long fibres the linear attenuation can be characterized by dissipation length L_γ related to an *e*-fold decrease of amplitude. The master equation (3.11) in a dispersive dissipative medium may be written (Brabec *et al.* 1991) as

$$i\frac{\partial f}{\partial \eta} + \frac{\partial^2 f}{\partial q^2} + \chi |f|^2 f = -i\gamma_0 f; \qquad \gamma_0 = \frac{L_\omega}{L_\gamma}. \tag{3.97}$$

The generalized soliton solution of eqn (3.97) in the form

$$|f|^2 = \frac{e^{-2\gamma_0\eta}}{\mathrm{ch}^2\left(\dfrac{q}{\Delta}\, e^{-2\gamma_0\eta}\right)} \tag{3.98}$$

predicts a decrease of the soliton's peak accompanied by an increase in its half-width. The typical value of dissipation length in silica fibres in the spectral range $\lambda_0 = 1.3{-}1.5\,\mu$m with losses $0.2\,\mathrm{dB\,km}^{-1}$ is $L_\gamma = 20{-}22\,$km.

The linear dissipation defined firstly for the bulk medium does not depend upon the mode of waveguide propagation. The non-linear losses of pumping wave power from Brillouin and Raman scattering may be independent of the modal structure of the trapped radiation. These losses occur due to light interaction with acoustic and optical phonons (molecular vibrations), respectively. However, the material dispersion of the medium itself can restrict the interaction of pumping and scattered waves.

In contrast, the interactions of two or three travelling waves belonging to waveguide eigenmodes depend upon the mode dispersion. This waveguide effect is shown in Section 3.5.1 to be responsible for formation of the phase shifts required for synchronous harmonic generation and the effective power loss of the pumping wave. The influence of material dispersion in Raman media on restriction of energy losses of the pumping wave produced by transformation to Stokes waves is discussed in Section 3.5.2.

3.5.1 *Doubling and tripling of frequencies in the waveguide*

The effectiveness of generation of the second and third harmonics of the pumping wave in the dispersive medium depends upon the phase relations between the interacting waves. If the interaction is not phase-matched, the generation of higher harmonics develops over a short distance, about several wavelengths; subsequently the accumulation of phase detuning between harmonics will stimulate the coupling of energy back to the pumping wave, and so on. To avoid these oscillations, compensation for the phase mismatch must be provided. The appearance of coherent wave–wave interactions produces effective conversion of pumping wave to new waves with doubled or tripled frequencies.

Second-order processes are normally dipole-forbidden in centro-symmetric silica-core optical fibres. However, self-organized phase-matched generation of the second harmonic in such fibre is allowed (Stolen and Tom 1987) by two pumping waves mixing with the harmonic wave by a third-order process characterized by d.c. polarization:

$$P_{\text{dc}} = \tfrac{3}{4}\, \chi^{(3)}(-2\omega, \omega, \omega)\, |E^*(2\omega)E^2(\omega)|\cos(\Delta\beta z). \qquad (3.99)$$

$\chi^{(3)}$ is the third-order non-linear susceptibility; $\Delta\beta$ is the detuning of longitudinal wavenumbers of interacting fibre modes:

$$\Delta\beta = \beta(2\omega) - 2\beta(\omega). \qquad (3.100)$$

The phase-matching periodicity is characterized by the coherence length (Laurell *et al.* 1988),

$$L_{\text{p}} = \frac{2\pi}{|\beta(2\omega) - 2\beta(\omega)|}. \qquad (3.101)$$

Another method of dispersion-dependent frequency doubling is connected with periodic modulation of the linear or non-linear properties of waveguide (Someth and Yariv 1972). The period Λ of the modulation is chosen so as to compensate for the phase mismatch between the interacting modes:

$$\Delta\beta = \frac{2\pi}{\Lambda}. \qquad (3.102)$$

Here the detuning $\Delta\beta$ is determined by (3.100). Modulation of the linear waveguide parameters for second-harmonic generation has been demonstrated in GaAs waveguides with surface corrugation (van der Ziel *et al.* 1976). Modulation of non-linear properties of the waveguide may be achieved, for example, by implementation of regions with alternating domain orientations in lithium niobate ferroelectric crystal. The alternating sign of the non-linear coefficient provided by such domain inversion can form the phase matching required for effective blue light generation in an $LiNbO_3$ channel waveguide by frequency doubling of laser diode radiation (Webjorn *et al.* 1989).

The optimal phase detuning between the pumping wave and its third harmonic can enhance the effectiveness of frequency tripling in the waveguide. Some segnetoelectrics such as $SrTiO_3$, described in the far IR and submillimetre ranges by non-linear susceptibility (1.6) with $\mathscr{E}_0$ about several thousands and $\chi_1 = -10^{-10}\,\mathrm{m^2V^{-2}}$, are important for third-harmonic generation. The self-consistent system of equations describing the interaction of pumping amplitude x_1 and its third harmonic x_3 in the rectangular waveguide may be written in the dimensionless form

$$\frac{\partial x_1}{\partial \eta} = -x_1^2 x_3 \sin \varphi; \tag{3.103}$$

$$\frac{\partial x_3}{\partial \eta} = x_1^3 \sin \varphi; \tag{3.104}$$

$$\frac{\partial \varphi}{\partial \eta} = \frac{(x_1^3 - 3x_1^2 x_3)}{x_3} \cos \varphi + 3\Gamma x_1^2 - 3\Delta x_3^2 - \delta. \tag{3.105}$$

Here φ is the phase detuning determined by the phases of interacting waves φ_3 and φ_1 and phase self-modulation $\delta\eta$ provided by the Kerr effect:

$$\varphi = \varphi_3 - 3\varphi_1 - \delta\eta. \tag{3.106}$$

Coefficients Γ, Δ, and δ are connected with the dispersion-dependent parameter γ_3:

$$\gamma_3 = \frac{3\beta_1}{\beta_3}; \tag{3.107}$$

$$\Gamma = \sqrt{\gamma_3}(2 - \gamma_3^{-1}); \qquad \Delta = \sqrt{\gamma_3}(2 - \gamma_3); \qquad \delta = M(\gamma_3^{-1} - 1). \tag{3.108}$$

The non-linear coefficient M is used for normalization of the distance η:

$$M = \frac{8c^2\beta_1^2}{\omega_1^2 \mathscr{E}_2 |E_1|^2}; \qquad \eta = \frac{3\beta_1 z}{M}. \tag{3.109}$$

Let us analyse the solutions of system (3.103–3.105) satisfying the following conditions:

$$x_1\big|_{\eta=0} = 1; \qquad x_3\big|_{\eta=0} = 0; \qquad \varphi\big|_{\eta=0} = \frac{\pi}{2}. \tag{3.110}$$

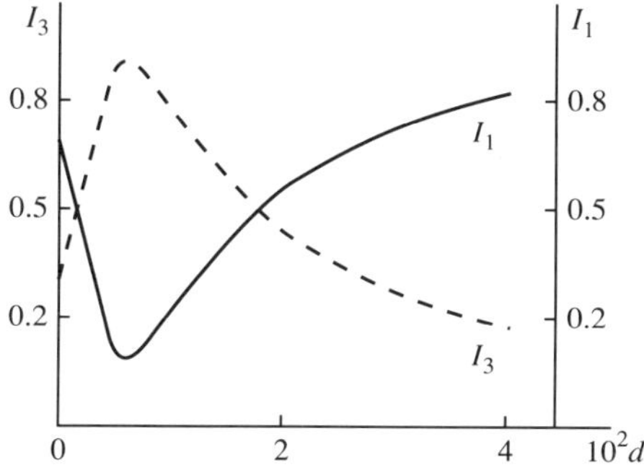

Fig. 3.14 Non-linear losses of pumping wave power due to third-harmonic generation in a waveguide. The dimensionless powers of pumping wave I (solid line) and third-harmonic I_3 (dashed line) are plotted against the dispersion-dependent detuning parameter $d = 1 - \gamma_3$ (3.107). The effectivities I_3 relate to the distance of maximum generation.

The system (3.103–3.105) has two conservation laws; in view of conditions (3.110) these laws may be written via the intensities $I_{1,3} = |x_{1,3}|^2$ of interacting waves:

$$I_1 + I_3 = 1; \tag{3.111}$$

$$\frac{3}{4}\left(\Gamma I_1^2 + \Delta I_3^2\right) + \frac{\delta}{2}I_3 - I_1\sqrt{I_1 I_3}\cos\varphi = \frac{3}{4}\Gamma. \tag{3.112}$$

The efficiency of third-harmonic generation as a function of the dispersion parameter $(d = 1 - \gamma_3)$ is shown in Fig. 3.14. The initial phase synchronism $(\gamma_3 = 1)$ leads to restriction of maximum available intensity of the third harmonic $I_3 = 0.31$ due to gradual violation of this synchronism in the course of propagation of the pumping wave produced by its phase self-modulation. The compensation of this effect by means of initial detuning of wavenumbers $\gamma_3 = 0.996$ is shown to provide the maximum effectiveness of generation $I_3 = 0.89$. Thus the small variations of the dispersion parameter γ_3 are responsible for a noticeable growth of non-linear losses of the pumping wave.

3.5.2 *Two-wave mixing and Raman generation in optical circuits*

The essential physics of wave interaction in a Raman medium may be considered in the framework of the two-field case. The formation of localized fields in this medium is connected with three energy levels $(0, 1, 2)$ with energies 0, $h\omega_{10}$, and $h\omega_{20}$, where the highest energy level

(1) is electric-dipole-coupled to the remaining two levels $(0, 2)$. The lower of those energy levels (0) is assumed to be an initially populated ground state. For simplicity we suppose the two transition dipole moments to be equal $\mu_{10} = \mu_{12} = \mu$.

Interaction of light with vibrations of molecular dipoles in a Raman medium provides the generation of Stokes waves with shifted frequency ω_s. Thus in silica fibre the first–Stokes light is primarily generated $440 \, \text{cm}^{-1}$ away from the pumping light. In the spectral range of abnormal dispersion $v_\omega < 0$ the group velocities of the pumping wave (v_g) and Stokes wave (v_s) fulfil the condition $v_s > v_g$. Thus the Stokes wave runs ahead and interaction of these waves in dispersive medium is shown to provide the formation of the two field localized wave.

To analyse the interaction of the pumping wave and the first-Stokes field, which is capable of potentially coupling levels (1–0) and (1–2) let us use the second-order Maxwell equation for plane waves travelling in the z-direction as the starting point:

$$\left(\frac{\partial^2}{\partial z^2} - \frac{1}{c^2} \frac{\partial^2}{\partial t^2} \right) \mathbf{E}(z, t) = \frac{4\pi}{c^2} \frac{\partial^2}{\partial t^2} \mathbf{P}(z, t). \qquad (3.113)$$

Here the polarization $\mathbf{P}$ may be written as

$$\mathbf{P}(z, t) = p\mathbf{E}(z, t) + ip[Q\mathbf{E}(z, t)e^{-i\delta t} - \text{c.c.}], \qquad (3.114)$$

$$Q = -ie^{i\delta t}\rho_{02}; \qquad \delta = \omega_p - \omega_s. \qquad (3.115)$$

ρ_{02} is the density matrix element relating to the (0–2) transition, $\mathbf{P}$ is the linear susceptibility of a medium connected with its dielectric permittivity $\mathscr{E} = 1 + 4\pi p$ and dependent upon the molecular density N,

$$p = \frac{2N\mu^2}{\hbar\omega_{10}}, \qquad (3.116)$$

and ω_p and ω_s are the frequencies of linearly polarized pumping and first-Stokes waves. Supposing their polarizations to be collinear, we can represent the field $E(z, t)$ as a scalar sum of these fields:

$$E(z, t) = \tfrac{1}{2} \left[E_p(z, t)e^{i(\beta_p z - \omega_p t)} + E_s(z, t)e^{i(\beta_s z - \omega_s t)} + \text{c.c.} \right] \qquad (3.117)$$

To simplify the mathematical treatment of the problem let us use two physically meaningful approximations. First we assume that we have many more molecules than photons; i.e. even if there is complete photon conversion from pump to first-Stokes, the molecular inversion, which tracks the number of converted photons, remains essentially in the ground state. The second approximation assumes that the optical medium is thin enough and the relative phases between interacting waves $\Delta\varphi = |k_p - k_s|z$ are small ($\Delta\varphi \ll 1$). With these suppositions in mind we can substitute the field (3.117) into eqn (3.113); using the method of slowly varying amplitudes yields the set of equations

$$\left(\frac{\partial}{\partial z} + \frac{1}{v_g}\frac{\partial}{\partial t}\right)E_p = \frac{QE_s}{L_p},$$ (3.118)

$$\left(\frac{\partial}{\partial z} + \frac{1}{v_g}\frac{\partial}{\partial t}\right)E_s = -\frac{Q^*E_p}{L_p}.$$ (3.119)

Here the length L_p is

$$L_p = \frac{h\omega_{10}cn_0}{4\pi N\mu^2\omega_p}$$ (3.120)

where $n_0 = \sqrt{1 + 4\pi p}$ is the refractive index of the medium.

To obtain a self-consistent set of equations it is necessary to supplement (3.118, 3.119) with the equation describing the temporal evolution of the density matrix element (3.115) of a two-level system (Allen and Eberly 1975):

$$\frac{\partial Q}{\partial t} = -\frac{e^{i\delta t}}{\omega_{10}}\left[\frac{\mu E(z,t)}{h}\right]^2.$$ (3.121)

The relaxation time in this model tends to infinity. Substituting the field $E(z,t)$ (3.117) into eqn (3.121) we obtain

$$\frac{\partial Q}{\partial t} = \frac{\mu^2}{2h^2\omega_{10}}E_pE_s^*.$$ (3.122)

It is worth defining a local time $q = t - zv_s^{-1}$ and a new variable $\theta(z,q)$ such that all the fields depend on z through θ. In addition we can determine

$$\frac{\partial\theta}{\partial q} = Q.$$ (3.123)

If a soliton solution of eqns (3.118, 3.119) does exist, it must be independent of z in the above-mentioned reference frame $q = t - zv_s^{-1}$: $\partial E_p/\partial z = \partial E_s/\partial z = 0$. Using the variable θ we may rewrite eqns (3.118–3.119) in the form:

$$\frac{\partial E_p}{\partial\theta} = \frac{E_s}{T_0}; \qquad \frac{\partial E_s}{\partial\theta} = \frac{E_p}{T_0};$$ (3.124)

$$T_0 = \frac{L|v_g - v_s|}{v_g v_s}.$$ (3.125)

The solutions of (3.124) are

$$E_{\mathrm{p}} = E_{p0}\cos\left(\frac{\theta}{T_0}\right); \qquad E_{\mathrm{s}} = E_{p0}\sin\left(\frac{\theta}{T_0}\right). \tag{3.126}$$

Here E_{p0} is the normalization constant: $E_{\mathrm{p}}^2 + E_{\mathrm{s}}^2 = E_{p0}^2$.

To find the hitherto unknown variable θ one can substitute the solution (3.126) into eqn (3.123):

$$\frac{\partial^2\theta}{\partial q^2} = -\frac{1}{t_0}\sin\left(\frac{\theta}{T_0}\right)\cos\left(\frac{\theta}{T_0}\right). \tag{3.127}$$

Here the time scale t_0 is

$$t_0 = \frac{2h^2\omega_{10}}{\mu^2 E_{p0}^2}. \tag{3.128}$$

Introducing the dimensionless variables F and x,

$$\theta = FT_0, \qquad q = t_0 x \tag{3.129}$$

We may find the first integral of eqn (3.127):

$$\left(\frac{\partial F}{\partial x}\right)^2 = \frac{t_0}{T_0}\cos^2 F. \tag{3.130}$$

The function F is given by integration of eqn (3.130):

$$\sin F = \mathrm{th}\left(\frac{q}{\sqrt{T_0 t_0}}\right). \tag{3.131}$$

Finally the solution (3.126) may be rewritten by means of (3.131) in an obvious form:

$$E_{\mathrm{p}} = \frac{E_{p0}}{\mathrm{ch}\left(\dfrac{q}{\sqrt{T_0 t_0}}\right)}; \qquad E_{\mathrm{s}} = E_{p0}\,\mathrm{th}\left(\frac{q}{\sqrt{T_0 t_0}}\right). \tag{3.132}$$

The characteristic time scale of the localized fields (3.132),

$$T_{\mathrm{c}} = \sqrt{T_0 t_0} = \frac{h\omega_{10}}{\mu^2 |E_{p0}|}\sqrt{\frac{c\,|v_{\mathrm{g}} - v_{\mathrm{s}}|}{v_{\mathrm{g}} v_{\mathrm{s}}}\frac{hn_0}{2\pi N\omega_{\mathrm{p}}}}, \tag{3.133}$$

is in inverse proportion to the pumping amplitude $|E_{p0}|$. The distributions (3.133) are narrowed by making the velocity v_{g} closer to v_{s}.

Thus we obtain the characteristic time scale T_{c} and the distributions of localized fields E_{p}^2 and E_{s}^2 formed due to interaction of the pumping wave and first-Stokes wave in a Raman medium (Fig. 3.15). The total

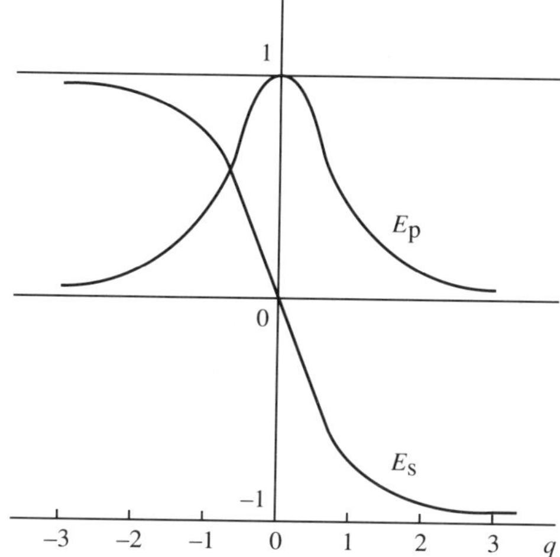

Fig. 3.15   Amplitude envelopes of a two-field Raman soliton ($E_\mathrm{p} =$ pumping pulse, $E_\mathrm{s} =$ first-Stokes pulse) are plotted against the normalized variable $q_1 = q(T_0 t_0)^{-1/2}$.

intensity of interacting fields (3.126) is conserved. Fig. 3.15 shows that a two-field soliton represents the appearance of pumping photons and the corresponding absence of first-Stokes photons, while in the region of the pump maximum a π phase shift arises in the electric field envelope of the first-Stokes wave.

This two-field example illustrates the methodology of the problem. The three-field case including the pumping wave, the first-Stokes, and the first-anti-Stokes waves as well as the two-pump four-field case are considered in a similar fashion.

3.6 Non-linear geometric optics of waveguides

The non-linear Schrödinger equation (3.11) has the special property that it is completely integrable for the inverse scattering transform. However, the level of mathematical art necessary for the use of this elegant method for analytic description of the evolution of intense pulses with limited path of evolution ($z \leqslant L_\omega$) proves to be extremely high. Some initial tendencies of such evolution may be outlined by means of the theory of perturbations and self-similar solutions that is well-known for Gaussian intensity profiles (Section 2.4.1). However, the numerical solution shows that the self-similar regime of evolution may be violated at limited distances $z = (2-3)L_\omega$. On the other hand, such limited, and even shorter, spans of the waveguide are important for the series of problems connected with high-bit-rate systems for fibre communication and optical processing. The development of non-linear geometric optics permits one to advance a theory for simple solution of these problems, which remain beyond of the scope of perturbation theory or self-similar solutions. The competition of non-linear and dispersive distortions of the localized wave field is characterized by the parameter χ in the master equation (3.11). Typical values of this parameter in non-linear pulse processes in optical fibres are usually large enough; thus, $\chi = 85$ in experiments with non-linear pulse compression (Grischkowsky and Balant 1982). The condition

$$\chi \gg 1, \tag{3.134}$$

often related to the supercritical regime $\chi > \chi_{cr}$, may be utilized for an essential simplification of equation (3.11) for non-stationary evolution of powerful self-localized fields in guiding systems. One may note that the 'trivial' geometric optics is often applied to waveguide propagation of 'weak' fields. Unlike the 'trivial' geometric optics deduced from Maxwell's equations, non-linear geometrical optics is based on the non-linear Schrödinger equation (3.11). This equation may be reduced to a set of coupled equations governing the dimensionless intensity W and the frequency function v:

$$W = |f|^2; \qquad v = \pm \sqrt{\frac{2}{\chi}} \cdot \frac{\partial s}{\partial q}. \tag{3.135}$$

Here the $+$ and $-$ signs coincide with the sign of the product $v_\omega \Delta n$. One of these equations relates to the real part of (3.11) and may be written as

$$\frac{\partial v}{\partial r} + v\frac{\partial v}{\partial q} \pm \frac{\partial W}{\partial q} + \frac{2}{\chi}\frac{\partial}{\partial q}\left(\frac{1}{\sqrt{W}}\frac{\partial^2 \sqrt{W}}{\partial q^2}\right) = 0. \tag{3.136}$$

The dimensionless distance τ is connected with the spatial variable η (3.11) by means of

$$\tau = \eta\sqrt{2\chi}; \tag{3.137}$$

the variable q is determined, as before, by (3.12).

The presence of the factor χ (3.134) in the denominator of the last term of eqn (3.136) permits one to neglect this term at least outside of the range of small amplitudes $W \to 0$. Omitting the last term in (3.136) and using an another equation obtained from the imaginary part of (3.11) without any new suppositios, one may write the set of dimensionless equations

$$\frac{\partial v}{\partial \tau} + v\frac{\partial v}{\partial q} \pm \frac{\partial W}{\partial q} = 0; \tag{3.138}$$

$$\frac{\partial W}{\partial \tau} + \frac{\partial (Wv)}{\partial q} = 0. \tag{3.139}$$

This set coincides formally with the equations of adiabatic one-dimensional flow of an ideal gas, the adiabatic index being equal to 2. The functions W and v may be considered as the density and velocity of such gas. The equations (3.138, 3.139) are the fundamental equations of non-linear geometric optics. Equation (3.138) represents the non-linear eikonal equation (3.139) in the transfer equation. The $+$ signs in (3.138) coincide with the signs of the products of non-linear and dispersion parameters χv_ω. In spite of the limitation $\chi \gg 1$, much relevant physics is still contained in the two coupled equations (3.138, 3.139).

The fundamental equations (3.138, 3.139) of non-linear geometrical optics describe the distribution of intensity W and frequency 'chirp' v along the pulse. The initial conditions to system (3.138, 3.139) are given by the envelopes of the input pulse

$$W\big|_{\tau=0} = W_0(q); \qquad v\big|_{\tau=0} = v_0(q). \tag{3.140}$$

In order to obtain the wide classes of exact analytic solutions of the system (3.138, 3.139) satisfying the initial conditions (3.140), it is convenient to transform this non-linear set by regarding W and v as independent variables and $\tau = \tau(W, v)$, $q = q(W, v)$ as dependent functions. It is notorious that such an operation transforms the non-linear system (3.138, 3.139) into a linear one in (W, v) space. Further, it is worth reducing the obtained set to one second-order differential equation in (W, v) space which expresses the quantities τ and q in terms of some function $\psi(W, v)$. This can be accomplished in two different ways by choosing the function ψ to satisfy exactly either the the transfer equation or the eikonal one. Let us consider the first way. In this case the functions $\tau = \tau(W, v)$ and $q = q(W, v)$ are introduced by the formulae

$$\tau = \pm \frac{\partial \psi}{\partial W}; \qquad q = 2v\tau \pm \frac{\partial \psi}{\partial v}. \tag{3.141}$$

Here and above the $+$ and $-$ signs relate to the $+$ and $-$ cases in eqns (3.138). However, the difficulties are not eliminated totally by this transformation; they are transferred to the initial conditions, which must now be written in an inverse form also. In order to surmount these difficulties it is necessary to choose in (W, v) space a special coordinate system suitable from the viewpoint of formulation of the boundary conditions. Let us introduce in a (W, v) space a spheroidal coordinate system (ξ, ε) convenient for the description of non-linear evolution of the pulse:

$$W = R^2(1 - \xi^2)(1 \pm \varepsilon^2); \qquad v = R\,\varepsilon\xi;$$

$$R^2 = \begin{cases} 1 \pm v_{\mathrm{m}}^2, & v_{\mathrm{m}}^2 < 1; \\ v_{\mathrm{m}}^2 \pm 1, & v_{\mathrm{m}}^2 > 1. \end{cases} \tag{3.142}$$

This co-ordinate system is shown in Fig. 3.16. v_{m} is the maximum value of the modulus of the function $v_0(q)$. In the special case $v_{\mathrm{m}}^2 = 1$, the standard spherical coordinate system becomes appropriate.

All the information about the pulse evolution is contained in the ψ function. The use of this function permits one to reduce the non-linear set (3.138, 3.139) via the Legendre transformation (3.141) to a linear Laplace equation:

$$\frac{\partial}{\partial \xi}\left[(1 - \xi^2)\frac{\partial \psi}{\partial \xi}\right] \pm \frac{\partial}{\partial \varepsilon}\left[(1 \pm \varepsilon^2)\frac{\partial \psi}{\partial \varepsilon}\right] = 0. \tag{3.143}$$

The initial conditions (3.141) in the (ξ, ε) system introduced on the surface $\varepsilon = \varepsilon_0$ have the form

$$\left.\frac{\partial \psi}{\partial \varepsilon}\right|_{\varepsilon = \varepsilon_0} = -2R\xi q_0(\mathrm{W}); \qquad \left.\frac{\partial \psi}{\partial \xi}\right|_{\varepsilon = \varepsilon_0} = -2v_{\mathrm{m}}q_0(W); \qquad \varepsilon_0 = \frac{v_{\mathrm{m}}}{R}.$$

$$\tag{3.144}$$

In the special case, when the initial modulation is absent ($v_{\mathrm{m}} = 0$), the surface ε_0 is a disc $\varepsilon = 0$ in (ξ, ε) space.

The procedure of exact analytical solution of non-linear geometric optics equations includes the following steps:

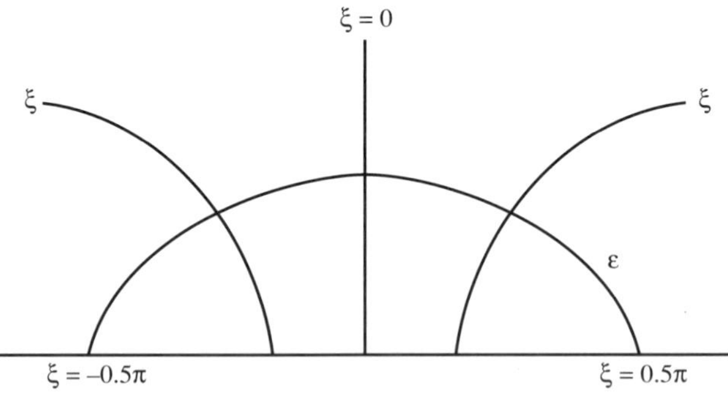

Fig. 3.16 The ellipsoidal ($\varepsilon = $ const.) and hyperboloidal ($\xi = $ const.) coordinate surfaces in (W, v) space.

(1) The function $\psi(\varepsilon, \xi)$ governed by (3.143) is written by means of spheroidal harmonics of the first (P_n) and second (Q_n) kinds. Thus, the following form of solution relates to the $+$ sign in (3.138) (Shvartsburg 1982):

$$\psi = \sum_n \{ Q_n(\xi) [A_n Q_n(\varepsilon) + B_n P_n(\varepsilon)] + P_n(\xi) [C_n Q_n(\varepsilon) + D_n P_n(\varepsilon)] \}$$

$$+ M_0 \ln [(1 - \xi^2) (1 \pm \varepsilon^2)] \tag{3.145}$$

The analogous solution related to the $-$ sign in (3.138), may be obtained from (3.145) by the change $\varepsilon \to i\varepsilon$. The coefficients (3.145) are determined by substitution of solution (3.145) into the boundary conditions (3.144).

(2) The inverse functions $\tau = \tau(\xi, \varepsilon)$; $q = q(\xi, \varepsilon)$ are calculated from the definitions (3.141), the function $\psi(\xi, \varepsilon)$ (3.145) being known. The solution ψ is built from the eigenfunctions of the Laplace equation (3.143).

It is instructive to emphasize that the solution of the wave problem under consideration is formally analogous to the determination of the potential function ψ in the space (ξ, ε) around some body via the values of ψ-function gradients on this body surface (3.144). This electrostatic analogy may be considered as a master key in the forthcoming calculations. Such a method permits one to obtain the wide classes of exact analytic non-stationary solutions describing pulse evolution up to the formation of very narrow localized field distributions. The possibility of unifying the analysis of non-stationary pulse processes for wide classes of envelopes attracts attention to this useful method.

Before discussing the conclusions from this lapidary recipe, let us illustrate its feasibility for pulse-shaping effects.

3.6.1 *Self-shaping of unmodulated pulses*

Pulses without frequency 'chirp' are characterized at the waveguide's input by the condition

$$v_{\mathrm{m}} = 0. \tag{3.146}$$

It is instructive to illustrate the method of non-linear geometric optics using the popoular model of localized wave pulse, the unmodulated (hyperbolic secant)^2 input pulse:

$$W|_{\tau=0} = \mathrm{ch}^{-2} q; \qquad v|_{\tau=0} = 0. \tag{3.147}$$

Let one fulfil the aforesaid operations;

1. An inverse initial profile may be written as

$$q|_{\varepsilon=0} = \frac{1}{2}\ln\frac{1+\xi}{1-\xi} = Q_0(\xi); \qquad \tau|_{\varepsilon=0} = 0. \tag{3.148}$$

2. The boundary conditions (3.144) combined with (3.145) yield the following values of coefficients in (3.145):

$$C_0 = B_1 = -2. \tag{3.149}$$

The other coefficients in (3.145) are equal to zero.

3. The functions q, τ obeying conditions (3.148) are

$$\tau = \frac{\varepsilon}{(1-\xi^2)(1\pm\varepsilon^2)} ; \qquad q = \frac{1}{2}\ln\frac{1+\xi}{1+\xi} \mp \frac{2\xi\varepsilon^2}{(1-\xi^2)(1\pm\varepsilon^2)}. \tag{3.150}$$

Restoration of the explicit formula for $W = W(\tau, q)$ and $v = v(\tau, q)$ via (3.150) is impossible. However, the evolution of the pulse may be analysed by means of these 'inverse' functions. The monotonic change of pulse slope may lead to the appearance of a region with an appreciable field gradient. In this region the derivative $\partial W/\partial q$ formally tends to infinity. Therefore, the conditions of quick intensity growth near the symmetric profile maximum may be written in the form

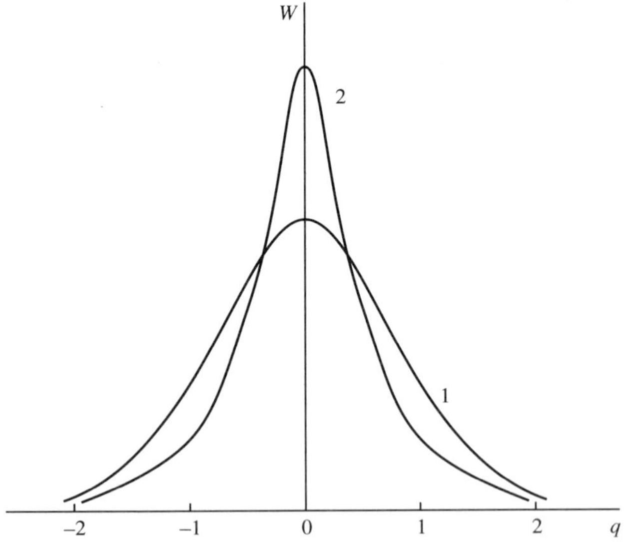

Fig. 3.17 Formation of a narrow spike-like intensity envelope $W(q)$ (curve 2) due to self-compression of a hyperbolic-secant pulse (curve 1) at the distance $\tau = 0.45$ close to the position of central peculiarity appear $\tau_c = 0.5$.

$$q = 0; \qquad \left.\frac{\partial q}{\partial W}\right|_{\tau} = 0.$$

This type of evolution leads to spike-like pulse formation in the case $\Delta n v_\omega > 0$ (Fig. 3.17). As well as such peculiarities, regions of quick intensity growth may occur at the wings of confined pulses as shown in Fig. 3.18 with $C < -1$,

$$q|_{\varepsilon=0} = \xi + \frac{1-\xi^2}{2C} \ln\frac{1+\xi}{1-\xi} ; \qquad \tau|_{\varepsilon=0} = 0. \qquad (3.151)$$

After the occurrence of such singularities the equations of non-linear geometric optics are no longer valid.

It is essential that the characteristic distance of these deformations development is of the order of magnitude $z \leqslant L_n$, which is shorter than the dispersive length L_ω from above condition $\chi = (\pi/2) L_\omega L_n^{-1} \gg 1$. Such non-stationary processes attract attention in view of problems of pulse compression preceding injection into the communication line. Considerable compression may be obtained due to pulse spectrum broadening and the posterior deformation of such pulse by passage through a grating pair or sodium vapour. The efficiency of this method depends upon the frequency distribution $v(q)$ along the pulse.

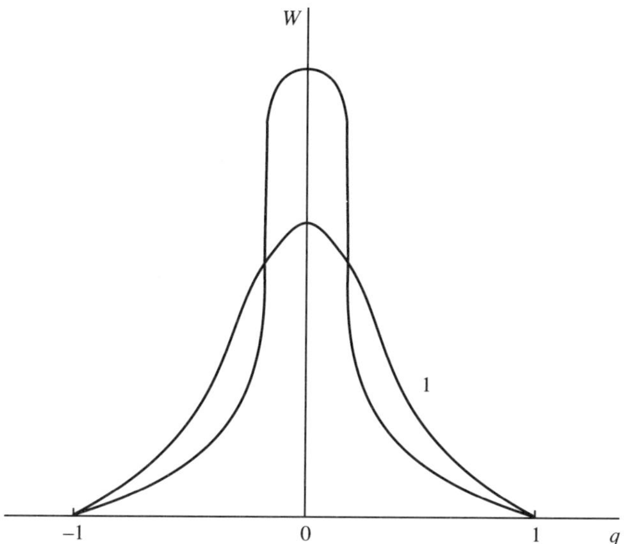

Fig. 3.18 Symmetrical self-steepening of the fronts of the pulse (3.151) (curve 1) localized in the interval $(-1 \leq q \leq 1)$ in the curve of evolution in a dispersive medium at the position $\tau = \tau_c$ related to 'spillover' formation.

Waveguide propagation proves to be an effective tool for the controlled spectral deformation. This deformation is accompanied by amplitude envelope tuning, and the simultaneous development of both these phenomena is manifested in Figs 3.19, 3.20.

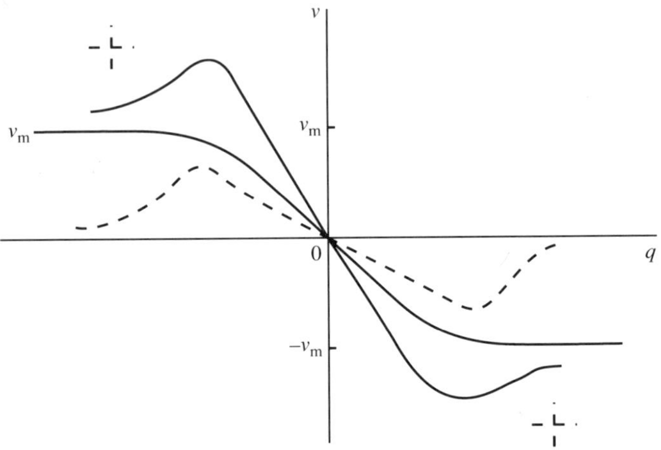

Fig. 3.19 Amplification of frequency 'chirp' (curve $+$) of the pulse $W_0 = \mathrm{ch}^{-2} q$ in the course of evolution due to addition of initial modulation $v_0 = v_m \,\mathrm{th}\, q$ and self-modulation (the dashed line).

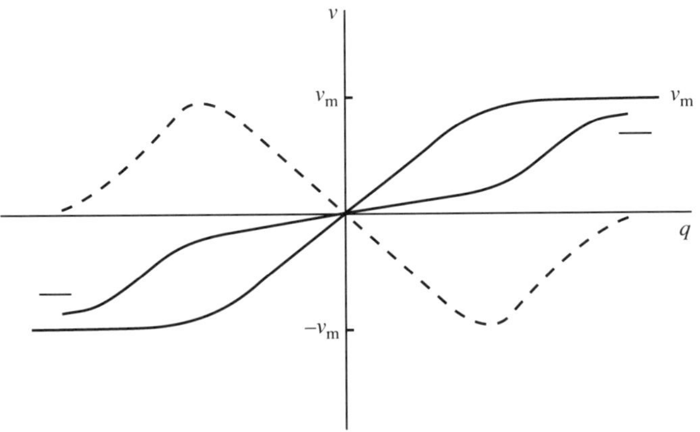

Fig. 3.20 Controlled demodulation of the pulse $W_0 = \mathrm{ch}^{-2} q$ at depth in a dispersive medium (curve $-$) due to subtraction of initial modulation $v_0 = v_m \,\mathrm{th}\, q$ from the frequency 'chirp' produced by non-linear self-modulation (the dashed line).

3.6.2 *Intensity-dependent evolution of modulated pulses*

The influence of the initial modulation of the pulse $(v_{\mathrm{m}} \neq 0)$ on the non-linear dynamics may also be analysed in the framework of the general system (3.138, 3.139). By slightly generalizing the previous algebra one can obtain in a straightforward way the coefficients in the solution (3.145), related, for instance, to the pulse

$$W_0 = \mathrm{ch}^{-2}q; \qquad v_0 = v_{\mathrm{m}}\mathrm{th}\,q. \tag{3.152}$$

The coefficients in (3.145), corresponding to the $+$ and $-$ cases, may be written by means of coefficient R (3.142):

$$C_0 = B_1 = 2\mathrm{i}R; \qquad M_0 = -v_m(+);$$
$$C_0 = B_1 = -2R; \qquad M_0 = -v_m(-). \tag{3.153}$$

The other coefficients are equal to zero, and the evolution of the modulated pulse (3.152) is described by the formulae

$$\tau = \frac{\varepsilon - \varepsilon_0}{R(1 - \xi^2)(1 \pm \varepsilon^2)}; \qquad q = \frac{1}{2}\ln\frac{1 + \xi}{1 - \xi} \mp \frac{2\xi\varepsilon(\varepsilon - \varepsilon_0)}{(1 - \xi^2)(1 \pm \varepsilon^2)}. \tag{3.154}$$

The influence of modulation is characterized here by means of the single parameter $\varepsilon_0 = v_{\mathrm{m}}R^{-1}$. Now it is obvious that the solution related to the unmodulated pulse may be obtained from (3.154) in the case $\varepsilon_0 = 0$.

The addition to initial modulation of frequency self-modulation, accumulated in the course of evolution, may accelerate or slow down the non-linear rebuilding of the intensity envelope. Thus, the spike-like pulse formation due to narrowing of the initially unmodulated pulse $W_0 = \mathrm{ch}^{-2}q$, $v_0 = 0$ is characterized by the distance of evolution $\tau_{\mathrm{cr}} = 0.5$; initial modulation of $v_0(q) = v_{\mathrm{m}}\mathrm{th}\,q$, $v_{\mathrm{m}} = 0.9$ (Fig. 3.19) with the signs of initial modulation and self-modulation being the same leads to this distance shortening to $\tau_{\mathrm{cr}} = 0.25$. On the other hand, opposite signs for these types of modulation (Fig. 3.20) may slow down the initial pulse deformation. In spite of the well-known tendency to non-linear pulse spreading in the waveguide with positive dispersion, this effect may be utilized for optical pulse stabilization, for instance, in the visible spectral range.

3.7 Pulse–pulse interactions in fibres

The problem of stabilization of a train of optical pulses in long communication cables attracts attention to the possibilities both of restoration of each pulse parameter in lossy media and of reducing of interaction of adjacent pulses. The proposed method of pulse recovery in dissipative fibres is connected with compensation of losses of one wave by means of Raman pump power produced by a counter-propagating wave flow with an appropriate shorter wavelength (Mollenauer *et al.* 1985). The characteristic spatial scale of this gain, L_R, can be evaluated as

$$L_R = \frac{16}{G_R P}. \tag{3.155}$$

Here G_R is the Raman gain coefficient ($G_R \leqslant 10^{-11}\,\mathrm{cm}\,\mathrm{W}^{-1}$ for fused silica), and P is the light power per cm^2 in the dielectric.

The periodic compensation of attenuation of information-carrying pulses may be provided by Raman pump light injected periodically into a fibre through the use of wavelength-dependent directional couplers spaced distance L apart. Because of decay of the pump power itself, Raman pump light launched at a given point can effectively cancel loss only over a limited amplication period L. The pulse energy W at some point z inside the fibre span L is connected with the pulse input energy W_0 by

$$W = W_0 \exp\left\{\frac{\alpha_s}{2}\left[L - 2z + \frac{L\,\mathrm{sh}\left[\alpha_p(z - L/2)\right]}{\mathrm{sh}(\alpha_p L/2)}\right]\right\}. \tag{3.156}$$

Here α_s and α_p are the fibre loss coefficients at signal and pump wavelengths, respectively. One may see that $W = W_0$ at three points, $z = 0$, $z = L/2$ and $z = L$. Thus, the recovery of the pulse energy occurs only at the points $L/2$, L; the value L in a real single-mode fibre system is of the order 30–50 km. Therefore, a system thousands of kilometres in overall length will contain many amplification periods L. The shape of the transmitted signal will change periodically.

In contrast to this effect based on the interaction of the wave pulse with a radiation flow, the frequencies of pulse and flow being different, it is worth discussing here some methods of tuning of interacting coherent pulses. The physical effects produced by these pulse–pulse interactions are shown to be different for short-time collision of counter-propagating pulses (section 3.7.1) and for gradual overlapping of co-propagating ones (Section 3.7.2).

3.7.1 *Phase effects in colliding pulses*

The momentary interaction of short pulses in the course of waveguide propagation in opposite directions has an influence, first of all, on their phases. To evaluate these perturbations we can generalize the model of non-reciprocal phase effects in counter-propagating waves given in Section 1.1. Thus, considering the collision of wave pulses with intensity envelopes $W_1(q_1)$ and $W_2(q_2)$ travelling in the TE_{01} mode in a parabolically-shaped fibre (1.62) with group velocity v_g, we may find the phase shift of pulses $W_{1,2}$ due to transitory overlapping of fields W_1 and W_2. Let the durations of interacting pulses be $2T_1$ and $2T_2$, respectively. Thus the distributions of phases $\varphi_{1,2}$ along the pulses after the interaction calculated from eqn (1.25) may be written as

$$\varphi_1 = \frac{\omega\chi v_\mathrm{g} T_2}{cn_0}\left[|E_1|^2 W_1(q_1) + |E_2|^2 \int_{-T_2}^{T_2} W_2(q_2)\,\mathrm{d}q_2\right], \quad (3.157)$$

$$\varphi_2 = \frac{\omega\chi v_\mathrm{g} T_1}{cn_0}\left[|E_2|^2 W_2(q_2) + |E_1|^2 \int_{-T_1}^{T_1} W_1(q_1)\,\mathrm{d}q_1\right]. \quad (3.158)$$

Unlike the CW radiation flows described by eqn (1.25), the non-reciprocal phase shifts of pulses (3.157, 3.158) depend on both peak amplitudes of interacting pulses and their dimensionless energies l_1 and l_2. The correlation of energies can change the sign of the phase shift for CW radiation by correlation of amplitudes E_1^2 and E_2^2.

Collisions of pulses belonging to different modes (Haelterman and Mestdagh 1987) is characterized by amplification of non-reciprocity of interaction due to mode dependence of the non-linear parameter χ_{0p}. Thus, the phase shifts of pulses with the same characteristic times T_0 propagating in the $\mathrm{TE}_{01}(\varphi_1)$ and $\mathrm{TE}_{02}(\varphi_2)$ modes may be written, in view of (3.15), as

$$\varphi_1 = \frac{\omega\chi v_\mathrm{g1} T_0}{4cn_0}\left[|E_1|^2 W(q_1) + \frac{5|E_2|^2}{16}\int_{-T_0}^{T_0} W_2(q_2)\,\mathrm{d}q\right], \quad (3.159)$$

$$\varphi_2 = \frac{5\omega\chi v_\mathrm{g2} T_0}{64cn_0}\left[|E_2|^2 W(q_2) + \frac{16|E_1|^2}{5}\int_{-T_0}^{T_0} W_1(q_1)\,\mathrm{d}q\right]. \quad (3.160)$$

The growth of the non-linear parameter χ close to cutoff in the TM_{01} mode (Section 1.4) promotes an increase of phase shifts of colliding TM_{01} polarized pulses.

Assuming, for example, a value of the non-linear length of phase self-modulation in the fibre of $L_\mathrm{n} = 20\,\mathrm{m}$, one can evaluate using (3.157) for picosecond pulses ($T_0 = 5\,\mathrm{ps}$) that the phase shift per pulse–pulse collision is about $10^{-4}\pi$. Thus 10^3–10^4 of these collisions may be sufficient for tuning of pulses in phase-sensitive systems such as non-linear couplers or switches (Section 1.5). Use of the same effects in non-linear polarizers and filters sensitive to phase shifts between the polarization components as small as $\varphi_0 = 1\text{--}2°$ (Fig. 1.29) can provide deep intensity modulation at the filter's output due to perturbation of φ_0 by means of only 10^2 pulse–pulse collisions.

The alterations of amplitudes of interacting counter-propagating pulses are ignored here. However, the longer interactions of partially overlapping pulses travelling in one direction are shown below to give rise to coupled amplitude–phase tuning.

3.7.2 *Interactions of co-propagating pulses*

The mutual influence of co-propagating pulses may be illustrated by means of two physically distinguished examples:

(1) Accumulation of perturbations in a train of pulses travelling in the same mode due to partial overlapping of their tails stipulated by the gradual spreading of pulses. These perturbations are important for choosing the distance between two adjacent signals in a fibre communication system. On the one hand, the use of short pulses results in increase of the total number of signals transmitted in a given time and a respective increase of information capacity of such system. On the other hand, the narrowing of these pulses promotes their diffusion, decreasing the distance between the adjacent pulses and providing interaction. The influence of these contradictory tendencies demands the use of an optimal distance between co-propagating pulses in the given span of fibre. To demonstrate these tendencies let us analyse the interaction of two pulses described at the waveguide's input $\eta = 0$ by function

$$f_0 = f(q - \Delta q) + f(q + \Delta q). \tag{3.161}$$

Here the distance between pulses' peaks at the point $\eta = 0$ is equal to $2\Delta q$. It is worth comparing the dynamics of hyperbolic-secant and Gaussian pulses, $f = f_{1,2}$:

$$f_1(q)|_{\eta=0} = \mathrm{ch}^{-1}\left(\sqrt{\frac{\chi}{2}}\,\frac{t}{T_s}\right), \tag{3.162}$$

$$f_2(q)|_{\eta=0} = \exp\left(-\frac{t^2}{2T_g^2}\right). \tag{3.163}$$

The peak amplitudes of both pulses f_1 and f_2 are equal to unity. It is convenient to compare the results of evolution of pulses (3.162) and (3.163), choosing their initial half-widths also to be equal. The characteristic times T_s and T_g must obey to condition

$$\frac{T_{\mathrm{g}}}{T_{\mathrm{s}}} = \sqrt{\frac{2}{\chi}} \, \frac{\mathrm{arch}\sqrt{2}}{\ln 2}.$$ (3.164)

Let us consider two partially overlapping pulses to be distinguished if the maximum value of intensity at any point between their peaks does not exceed half of their peak intensities. Determining the minimum distance between input pulses, $2\Delta q_0$, from this condition applied to the diffusing pulses at fibre's output, we can show by means of computer simulation of the non-linear Schrödinger equation with initial conditions (3.161–3.163) the existence of optimal values of parameters q and Δq_0 for each type of pulse for given length of fiber span (Fig. 3.21). These parameters may be recommended for elaboration of non-linear fiber communication circuits with given length (Crosignani *et al.* 1981).

(2) Self-trapping of co-propagating pulses travelling in different modes based on non-linear compensation of modal dispersion. Let us consider hyperbolic-secant pulses launched simultaneously in a parabolically-shaped waveguide in the TE_{01} and TE_{02} modes. As was pointed out by Hasegawa and Kodama (1981), the interaction of such pulses f_1 and f_2 with the same duration T_0 is described by the set of equations

$$\mathrm{i}\left(\frac{\partial f_1}{\partial \eta} - \delta \frac{\partial f_1}{\partial q}\right) + \frac{\partial^2 f_1}{\partial q^2} + \chi_0 [\,|f_1|^2 + |f_2|^2\,]f_1 = 0;$$ (3.165)

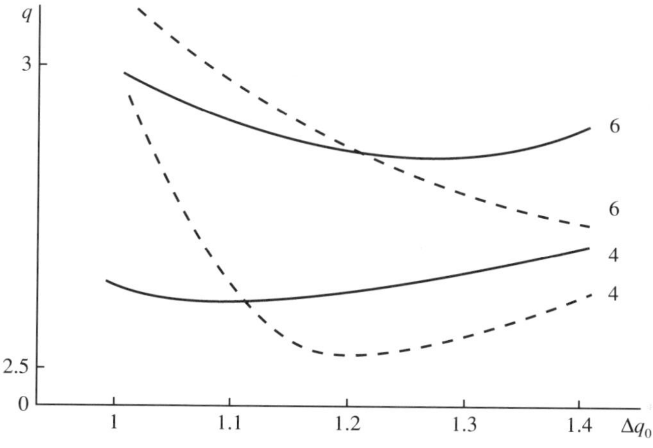

Fig. 3.21   Dependence of scales Δq_0 on q (3.161) for hyperbolic-secant (solid lines) and Gaussian (dashed lines) pulses to be distinguished at the output of the waveguide; the curves relate to the distances of propagation $\eta = 4$ and 6 respectively.

$$i\left(\frac{\partial f_2}{\partial \eta} + \delta \frac{\partial f_2}{\partial q}\right) + \alpha \frac{\partial^2 f_2}{\partial q^2} + \chi_0 [\, |f_2|^2 + \tfrac{5}{8} |f_1|^2] f_2 = 0. \qquad (3.166)$$

Here

$$\delta = \frac{L_\omega}{2T_0}\left(\frac{1}{v_{g2}} - \frac{1}{v_{g1}}\right); \qquad \alpha = \frac{v_{2\omega}}{v_{2\omega}}\left(\frac{v_{1g}}{v_{2g}}\right)^2. \qquad (3.167)$$

$v_{g1,2}$ are the group velocities in the TE_{01} and TE_{02} modes respectively, L_ω is the dispersive length for the pulse f_1,

$$q = T_0^{-1}\left[t - z\left(\frac{1}{v_{g1}} - \frac{1}{v_{g2}}\right)\right], \qquad (3.168)$$

and the coefficients (3.15) are taken into account.

Computer simulation of the dynamics of pulses $f_1|_{\eta=0} = f_2|_{\eta=0} = \mathrm{ch}^{-1}(tT_0^{-1})$ with $T_0 = 1\,\mathrm{ps}$, $\lambda_0 = 1.5\,\mu\mathrm{m}$ in the fibre (1.62) with parameters $a = 50\,\mu\mathrm{m}$, $\Delta = 0.03$ illustrates the difference of the behaviour of these pulses in the cases of weak ($\chi_0 = 0.5$) and strong ($\chi_0 = 4$) non-linearities. The case $\chi_0 = 0.5$ relates to the running of pulses with group velocities $v_{g1,2}$ close to those determined from the linear theory. In contrast, the strong non-linearity ($\chi_0 = 4$) provides compensation of model dispersion and self-trapping of pulses f_1 and f_2 propagating with coinciding velocities. The similar trapping of solitons in a slightly birefringent optical fibre may be a possibility for elaboration of ultra-fast optical logic gates (Islam 1989).

3.8 Shaping of non-soliton pulses in guiding systems

The dynamics of both soliton and non-soliton pulses analysed above was characterized by the conservation of the pulse's symmetry: $W(q) = W(-q)$. This effect was obtained for fixed polarization of the TE_{0p} travelling modes containing only one polarization component, the pulse duration $2T_0$ being much greater than the relaxation time of the non-linear response of the medium τ_n. Some small violations of initial pulse symmetry in the model under discussion could arise due to power dependence of group velocity (Fig. 3.2) or second dispersion (Fig. 3.3). These complicated pulse deformations, e.g. the splitting of the second-order soliton's envelope (Fig. 3.6), were shown to develop symmetrically.

However, the general tendency to controlled shaping and shortening of optical pulses up to femtosecond limits (Spielmann *et al.* 1991) points out the possibilities of splitting and asymmetrical deformations of pulse profiles. Power-dependent pulse shaping including splitting may be realized by phase and polarization modulation of radiation in a waveguide with light-induced (Section 1.1) or geometric (Section 1.4) birefringence. Using of an analyser at the output of such a waveguide can provide controlled transformation of amplitude envelopes. These effects, important for shaping of non-soliton pulses described in Section 3.3.2, will be analysed in Section 3.8.1. The influence of final relaxation time on the asymmetrical steepening of primarily symmetrical short pulses is illustrated in Section 3.8.2.

3.8.1 *Non-linear polarization phenomena in the shaping of wave pulses*

The non-linear rotation of polarization ellipses described in Section 1.1 for CW radiation manifests itself also for pulsed radiation. However, the influence of amplitude–phase modulation distributed along the input pulse will provide the formation of a related distribution of turning angles along the envelope at the waveguide's output. Thus, choosing for simplicity the span of a capillary waveguide with length Z_0 (Section 1.5.2), corresponding to the turning angle $\theta = \pi/2$ for peak amplitude E_0 and phase shift between the polarization axes at the pulse maximum $\varphi = \pi/2$, one can write this distribution of angles $\theta = \theta(q)$ along the pulse with intensity profile $W_0(q)$ in the form

$$\theta(q) = \frac{\pi}{2} W_0(q) \, sin\left[\frac{\pi}{2} \varphi_0(q)\right]. \tag{3.169}$$

The case of purely phase modulation ($W_0 = 1$) of CW radiation was analysed in Section 1.5. To illustrate the effects of self-narrowing and self-flattening of the pulse, let us consider for simplicity purely amplitude modulation ($\varphi_0 = 1$). According to (3.169) the turning angle decreases from the peak of amplitude ($W_0 = 1$) to the edges of the pulse. The non-linear filtration of transmitted radiation described by the coefficient (1.268) can provide, for example, better transmittivity for the peak amplitude ($\theta = \pi/2$), while a decrease of coefficient K (1.268) for the smaller turning angles ($\theta < \pi/2$) indicates a narrowing of the pulse passing through the filter (Fig. 3.22a). In contrast, a decrease of transmittivity near the peak amplitude accompanied by weak modulation of the coefficient K at the edges of the envelope can give a rise to flattening of the pulse and even to the pulse's splitting (Fig. 3.22b).

The parameter μ describing the ratio of the axes of the polarization ellipse of the input radiation was considered hitherto to remain constant. The alteration of this parameter $\mu = \mu(q)$ can stimulate the

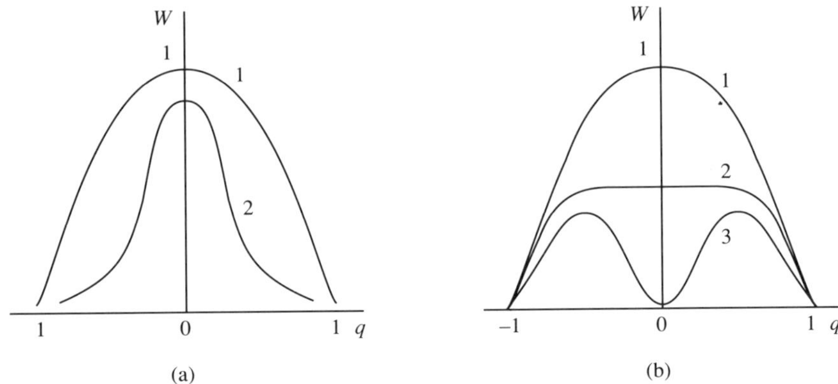

Fig. 3.22   Transformation of an initially parabolic pulse $W_0 = 1 - q^2$ (curve 1) of an elliptically polarized wave due to passage through the non-linear filter (1.268). (a) Pulse narrowing (curve 2); (b) pulse flattening (curve 2); self-stratification (curve 3).

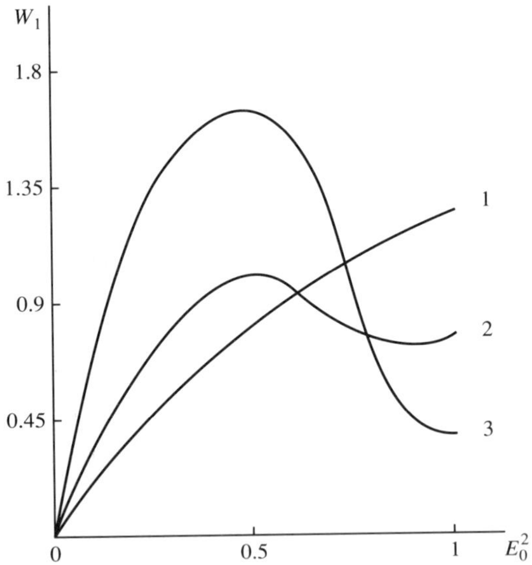

Fig. 3.23   Monotonic and non-monotonic dependences of normalized transmitted intensity $W_1 = E_1^2 E_c^{-2}$ of a non-linear polarization filter (1.268) upon the normalized intensity of the input wave $E_0^2 = 2\mu E^2 E_c^{-2} \pi^{-1}$, $E_c^2 = \lambda (\pi z \sin \varphi_0 \chi B)^{-1}$ (see Section 1.1). The curves 1, 2, and 3 relate to the values $\mu = 0.8$, 0.5, and 0.25, respectively.

modulation of transmittivity of this tunable filter producing, in particular, pulse splitting (Fig. 3.23). The phase shift giving this non-linear filtration can be formed by light-induced birefringence both in a capillary waveguide (Section 1.1) and a strip fibre (Section 1.3). Combination of these types of modulation is a useful tool for shaping of soliton and, especially, complicated non-soliton pulses.

3.8.2 *Non-stationary formation of asymmetrical localized fields*

The evolution of a primarily symmetrical powerful pulse in a dispersionless medium was shown to result in an asymmetrical shift of pulse maximum and steepening of one of the fronts (Fig. 3.1). However, the pulse's peak amplitude remains constant in the course of such evolution. Similar tendencies may develop in dispersive media also, the travelling pulse being ulrta-short. Such distortions of the initial pulse can arise for picosecond pulses travelling in liquid dielectrics and for femtosecond pulses in silica fibres. The main similar tendencies for both these non-stationary processes are outlined here.

The above analysis relates to the case of comparatively long pulses, when the pulse duration $2T_0$ is much longer than the time scale τ_n for establishment of non-linearity. This condition is fulfilled in glass waveguides for pulses up to sub-picosecond. On the other hand, the strong non-linearities intrinsic to some liquid dielectrics such as CS_2 are characterized by the value $\tau_n = 1\text{--}2\,\text{ps}$. Therefore, non-linear evolution of sub-picosecond pulses in the capillaries filled with such liquids is produced by the gradual establishment of non-linearity. Here the different parts of the pulse are travel through the medium with the ingtantaneous values of non-linear perturbation of the refractive index. The non-stationary process of this perturbation, Δn relaxation, may be described by means of the model equation (Stageman *et al.* 1988)

$$\frac{\partial \Delta n}{\partial t} + \frac{\Delta n}{\tau_n} = \frac{n_2 |\mathbf{E}|^2}{\tau_n}. \tag{3.170}$$

The short time scale $(\tau_n \ll 2T_0)$ relates to the case $\Delta n = n_2 |\mathbf{E}|^2$ discussed above. The opposite case $(\tau_n \gg 2T_0)$ leads in the framework of non-linear geometric optics to the eikonal equation different from eqn (3.138):

$$\frac{\partial v_1}{\partial \tau_1} + v_1 \frac{\partial v_1}{\partial q} \pm W = 0. \tag{3.171}$$

The transfer equation (3.139) remains valid after the replacement

$$\tau \rightarrow \tau_1; \qquad v \rightarrow v_1$$

$$\tau_1 = \frac{z}{L_0} \;; \qquad v_1 = \frac{L_0}{L_\omega} \frac{\partial s}{\partial q} \;; \qquad L_0 = L_n \sqrt{\tau_n/2T_0}. \qquad (3.172)$$

The role of non-linear phenomena is impeded here $(L_0 \gg L_n)$; thus the condition of utilization of non-linear geometrical optics $(x \gg 1)$ is modified:

$$\frac{2T_0\chi}{\tau_n} \gg 1. \qquad (3.173)$$

It is instructive to illustrate the influence of non-linearity relaxation by means of a 'bell-like' pulse $W_0 = \mathrm{ch}^{-2}q$. Thus, the exact analytical solution of a non-linear set of equations (3.171) and (3.139), relating to the + sign in (3.138), may be written as

$$W = \left\{ \mathrm{ch}^2\left(q - \frac{v_1\tau_1}{2} \right) - \frac{\tau_1^2}{2} \right\}^{-1},$$

$$v_1 = \frac{\tau_1}{2}\left\{ 1 - \mathrm{th}\left(q - \frac{v_1\tau_1}{2} \right) \right\}. \qquad (3.174)$$

The solution (3.174) describes the growth of the maximum and the compression of the pulse accompanied by peak displacement towards

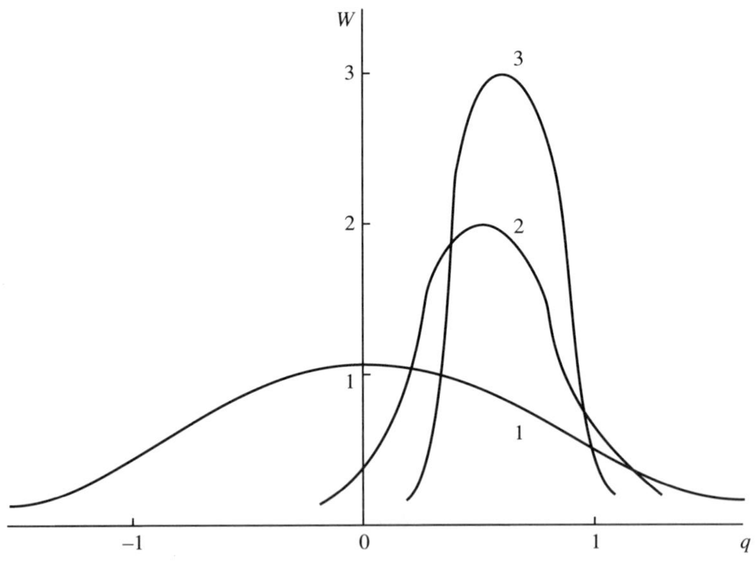

Fig. 3.24   The self-confinement of pulse a $W_0 = \mathrm{ch}^{-2}q$ accompanied by loss of symmetry in a medium with inertial non-linearity; the curves 1, 2, and 3 relate to the values $\tau_1 = 0$, 0.5, and 1.1, respectively.

the ascending part of the envelope (Fig. 3.24); the pulse's half-width diminishes at four times at the distance $\tau_1 = 1.1L_0$. The frequency 'chirp' increases along the pulse monotonically.

These examples show the difference in trends of pulse evolution in two opposite models, $2T_0 \gg \tau_n$ and $2T_0 \ll \tau_n$. In view of the short duration, high peak amplitude, and steepened front, these asymmetrically shaped envelopes seem to be candidates for the physics and applications of ultra-short pulses.

CONCLUSION

1. The sensitivity of non-linear evolution of solitary waves to their initial amplitude-phase envelopes provides the existence of both self-compression and self-lamination regimes of tuning. Side-by-side with several examples of monotonic self-compression of input pulses analysed in Sections 3.3.2 and 3.6.1 we can mention an illustration of periodical self-lamination of initially smooth envelopes of the second order solitons given in Section 3.3.1. However this oscillating type of evolution is not an exceptional case of pulse transformations.

To emphasize again an influence of fine details of initial conditions on the pulse dynamics let us show that even the small perturbations of these conditions can change radically the regime of evolution from, e.g., formation of spike-like pulse to the envelope stratification or vice versa. Thus, the self-constriction of unmodulated solitary wave with the parabolic intensity profile

$$W\big|_{\eta=0} = \begin{cases} 1 - q^2; & q^2 \leqslant 1; \\ 0; & q^2 \geqslant 1; \end{cases} \qquad V_{\eta=0} = 0 \qquad (3.175)$$

is known to provide the monotonous growth of its maximum and steepening of fronts (Zakharov *et al.* 1980). At the same time some distortions of this profile, its maximum and duration being kept constant

$$W\big|_{\eta=0} = \begin{cases} (1 - q^2)(1 - q^2 + q^4); & q^2 \leqslant 1; \\ 0; & q^2 \geqslant 1; \end{cases} \qquad V\big|_{\eta=0} = 0 \qquad (3.176)$$

stimulate the gradual decrease of steepness at the periphery of profile $W(q)$ in the course of propagation and formation of bend points determined by equations

$$\frac{\partial W}{\partial q} = 0; \qquad \frac{\partial^2 W}{\partial q^2} = 0; \qquad (3.177)$$

The computer simulation of solution of non-linear Schrödinger equation (3.11) with the initial conditions (3.176) shows, that the self-confinement of central peak is accompanied by the development of two new maxima separated from this peak by a pair of new minima (Fig. 3.25). The distances from the pulse centre to these minima are about $(0.3 \div 0.5)v_g T_0$. Such large scale self-stratification effect, which may become a source of errors in optical information systems, draws an additional attention to the methods of shaping of input signals in such systems.

2. The non-linear stabilization of solitary pulses is known to be impeded due to wave power dissipation. However, the decrease of density of power flow in the lossy fiber may be compensated by means of gradual diminiation of the guiding area

$$a(z) = a_0 \exp(-z/L_\gamma) \qquad (3.178)$$

here a is the radius of guiding area, L_γ is the dissipation length. Therefore the peak amplitude and the width of the lowest order soliton (3.30) propagating in this fibre could be made invariant, if the fibre is made axially non-uniform according to (3.178). Since such 'geometric' stabilization does not need any additional energy sources, this method is interesting for some spans of waveguide with narrowing core, restricted, e.g., by cutoff conditions.

3. Discussing in this Chapter the non-linear dynamics of wave pulses in guiding systems we used the polarization vectors determined by the

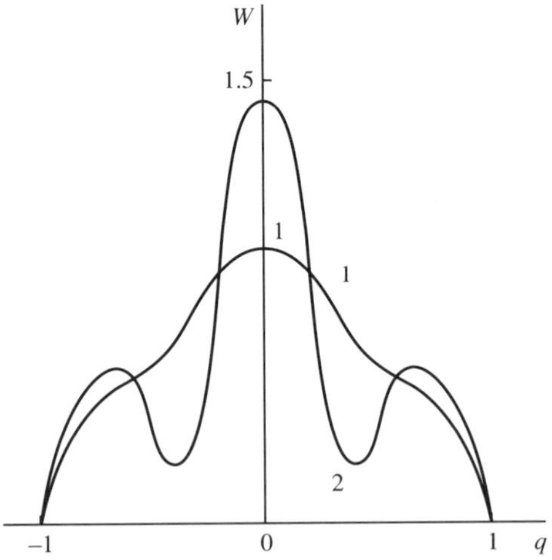

Fig. 3.25   The self-lamination of pulse W(q) (3.176) in the non-linear waveguide; the curves 1 and 2 relate to the values $\eta = 0$ and $\eta = 0.35$ respectively.

eigenmodes of trapped monochromatic cw radiation. These vectors are known to depend upon the frequency; therefore the spectral envelopes of localized pulse with final spectral bandwidth must be characterized by different polarization structures. Hence it appears, that the traditional model of fixed polarization vectors depending upon the carrying frequency and ignoring the final spectral bandwidth provides only an approximated description of pulse evolution. To clarify the feasibility of this approximation we must reconsider the derivation of non-linear Schrödinger equation in order to take into account the derivatives of mode polarization vectors $\vec{j}(\omega)$ with respect to frequency ω.

For simplicity let us discuss from this viewpoint the pulse travelling in TE_{01} mode in a circularly-shaped waveguide with a parabolic profile of refractive index (1.62). Revealing the $\vec{j}(\omega)$ dependence we may express an electric field of this pulse E_θ by means of the corresponding function Ψ_{01} (1.67) written in a form:

$$E_\theta = E_0 \Psi_{01}; \qquad \Psi_{01}(\omega, \rho) = \sqrt{2}\, j_\theta(\omega) \frac{\rho}{a} \exp\left(-\nu\rho^2/2a^2\right) \quad (3.179)$$

Here E_0 is the maximum value of electric field, $\sqrt{2}$ is the normalization constant (1.67),

$$j_r = 0, \qquad j_z = 0,\, j_\theta = \nu; \qquad \nu = a\omega n_0\sqrt{2\Delta}/c \quad (3.180)$$

Parameters a and Δ are determined by the waveguide model (1.62). While keeping (3.179) we can examine the dispersion of mode vectors simultaneously with the dispersion of group velocity v_g.

To accomplish this programme let us combine the expansion of dielectric permittivity $\varepsilon(\omega)$ given in Section 3.1 with an analogous expansion of electric field of the pulse envelope $f(z, t)$

$$E_\theta = E_0 \left[f(z, t)\Psi_{01}(\omega, \rho) + i\, \frac{\partial f(z, t)}{\partial t} \frac{\partial \Psi_{01}(\omega, \rho)}{\partial \omega} + \ldots \right] \quad (3.181)$$

Proceeding in the same fashion, that in the course of derivation of master equation (3.11) we obtain the modified non-linear Schrödinger equation

$$i\, \frac{\partial f}{\partial \eta} + \frac{\partial^2 f}{\partial q^2} + \chi |f|^2 f = Q; \quad (3.182)$$

$$Q = i\chi_1 \frac{\partial}{\partial q}\left(|f|^2 f\right) + i\chi_2 f\, \frac{\partial |f|^2}{\partial q}; \quad (3.183)$$

$$\chi_1 = \frac{\chi}{4\omega T_0}\left(1 - \frac{\omega n_\omega}{n_0}\right); \qquad \chi_2 = \frac{\chi}{\omega T_0}\left(1 + \frac{\omega n_\omega}{n_0}\right); \qquad n_\omega = \frac{\partial n_0}{\partial \omega};$$

$$(3.184)$$

Here χ_1 and χ_2 are the dimensionless power-dependent coefficients, the term Q (3.183) arises due to consecutive use of an expansion (3.181). These terms could be neglected in the analysis of long pulses containing many wavelengths ($\omega T_0 \gg 1$). The traditional form of master equation with Q = 0 ignoring the dispersion of mode vectors is justified in this limit.

In contrast, the role of mode vectors dispersion is increasing for narrow pulses containing several wavelengths ($\omega T_0 \gg 1$). The effective scales of radial confinement of spectral harmonics $\rho_0 = a\nu^{-1/2}$ determined from (3.179) prove to be different due to wide spectra of these pulses. The concept of fixed mode vectors determined for cw radiation is violated in the limit of ultra-short pulses. Hence it appears, that the influence of term Q in Eqn (3.182) connected with these effects on the dynamics of such localized fields may become decisive.

References

Agrawal, G. P. (1989). Non-linear fiber optics. Academic Press, New York.

Allen, L. and Eberly, I. H. (1975). *Optical resonance + two-level atoms.* Wiley, New York.

Anderson, D. (1983). Variational approach to non-linear pulse propagation in optical fibers. *Physical Review* **A27**, 3135–45.

Antoniuk, A. D., D'yacherko, A. G., Ekzhanov, R. I., Karlov, N. V., Sisakyan, I. N., Skorik, V. A. *et al.* (1989). Controlling millimeter- and submillimeter-range radiation as it interacts with a semiconductor plasma. *Soviet Physics Doklady* **34**, 624–6.

Avrutsky, I. A. and Sychugov, V. A. (1990). Optical bistability due to excitation of a non-linear corrugated waveguide. *Quantum Electronics* [in Russian] **17**, 933–7.

Bartlett, H. O., Lohmann, A. W., Freude, W., and Grau, G. K. (1983). Mode analysis of optical fibers using computer-generated matched filters. *Electronics Letters* **19**, 247–9.

Berdaque, S. and Facq, P. (1982). Mode division miltiplexing in optical fibers. *Applied Optics* **21**, 1950–5.

Bloembergen, N. and Pershan, A. S. (1962). Light waves at the boundary of non-linear dialectric. *Phys. Rev.*, **128**, 606–14.

Brabec, T. F., Spielmann, Ch., Wintner, E., and Budil, M. (1991). Effects of dispersion in additive-pulse mode-locked lasers. *Journal of the Optical Society of America* **B8**, 1818–23.

Brinca, A. L. and Mendonca, J. T. (1979). Refraction into a non-linear medium: wave profiles and hysteresis. *Physica Scripta* **19**, 545–9.

Chandra, P., Coldren, L. A., and Strege, K. E. (1981). Refractive index data from GaInAsP films. *Electronics Letters* **17**, 6–7.

Crosignani, B., Papas, C. H., and Di Porto, P. (1981). Coupled mode theory approach to non-linear pulse propagation in optical fibers. *Optics Letters* **6**, 6–13.

Dallas, W. I. and Lohmann, A. W. (1972). Quantization and other non-linear distortions of the hologram transmittance. *Optics Communications* **5**, 78–81.

Doran, N. I. and Blow, K. I. (1983). Solitons in optical communications. *IEEE Journal of Quantum Electronics* **QE-19**(12), 1883–7.

Ferber, F. S. and Marburger, J. H. (1976). Theory of non-resonant multistable optical devices. *Applied Physics Letters* **28**, 731–5.

Garitchev, V. B., Golub, M. A., Karpeev, S. V., Krivoshlykov, S. G., Petrov, N. I., Sisakyan, I. N. *et al.* (1985). Experimental investigation of mode coupling in a multimode graded-index fibre caused by periodic microbends using computer-generated spatial filters. *Optic Communications*, **55**, 403–5.

Gibbs, H. M. (1985). *Optical bistability: controlling light by light.* Academic Press, New York.

Gibbs, H. M., McCall, S. L., and Venkatesan, T. N. C. (1976). Differential gain and bistability using a sodium-filled Fabry-Perot interferometer. *Physical Review Letters* **36**, 1135–8.

Grau, G. K., Leminger, O. C., and Sauter, E. G. (1980). Mode excitation in parabolic index fibers by Gaussian beams. *Archiv för Electronic und Ubertragungstechnik* **34**, 259–65.

Grischkowsky, D. and Balant, A. C. (1982). Optical pulse compression based on enhanced frequency chirping. *Applied Physics Letters* **41**, 1–3.

Haelterman, M. and Mestdagh, D. (1987). Non-linear propagation in multimode optical fibers. *Optics Communications* **63**, 205–9.

Hasegawa, A. and Kodama, Y. (1981). Signal transmission by optical solitons in multimode fiber. *Proceedings of the IEEE* **69**(9), 1145–50.

Ichioka, Y. and Lohmann, A. W. (1972). Interferometric testing of large optical components with circular computer holograms. *Applied Optics* **11**, 2597–602.

Janke, E., Emde, F., and Losch, F. (1960). *Tafeln Höherer Funktionen, Seghste Auflage.* B. G. Teubner Verlagsgesellschaft. Stuttgart.

Islam, M. N. (1989). Ultrafast all-optical logic gates based on soliton trapping in fibers. *Optics Letters* **14**, 1257–9.

Jensen, S. M. (1982). The non-linear coherent couplers. *IEEE Journal of Quantum Electronics*, **QE-18**, 1580–3.

Jordan, I. A., Hirsch, P. H., Lesen, L. B., and Rooy, D. L. (1970). Kinoform lenses. *Applied Optics* **9**, 1883–9.

Kaplan, A. E. (1980). Bistable reflection of light by an electro-optically driven interface. *Applied Physics Letters* **38**, 67–9.

Kaplan, A. E. (1981). Optical bistability that is due to mutual self-action of counterpropagating beams of light. *Optics Letters* **6**, 360–2.

Kaplan, A. E. (1983). Light induced non-reciprocity, field invariants and non-linear eigenpolarizations. *Optics Letters* **8**, 560–2.

Karlov, N. V., Shvartsburg, A. B., and Sissakyan, E. V. (1985). Optical bistability produced by resonant Joule heating of semiconductors. *Applied Physics* **B35**, 77–81.

Koch, S. W., Schmidt-Rink, S., and Haug, H. (1981). Calculation of the intensity-dependent changes of the index of refraction in GaAs. *Solid State Communications* **38**, 1023–6.

Laurell, F. and Arvidsson, G. (1988). Frequency doubling in TI; MgO: LiNbO$_3$ channel waveguide. *Journal of the Optical Society of America* **B5**, 292–5.

Maker, D. P., Terhune, R. W., and Savage, C. M. (1964). Non-linear polarization of light in the liquid dielectric. *Phys. Rev. Lett.*, **12**, 507–10.

Marcuse, D. (1974). *Theory of dielectric optical waveguides.* Academic Press, New York and London.

Mayer, A. and Keilmann, F. (1986). Far-infrared non-linear optics. *Physical Reviews* **B33**, 6954–68.

Menyuk, C. R. (1989). Pulse propagation in an elliptically birefringent Kerr medium. *IEEE Journal of Quantum Electronics* **QE-25**, 2674–82.

Mollenauer, L. F., Evangelides, S. G., and Gordon, J. P. (1991). Wavelength division multiplexing with solitons in ultra-long distance transmission using Lamped amplifiers. *Journal of Lightwave Technology* **9**, 362–7.

Mollenauer, L. F., Stolen, R., and Islam, M. N. (1985). Experimantal demonstration of soliton propagation in long fibers: loss compensated by Raman gain. *Optics Letters* **10**, 229–31.

Petit, R. (1980). *Electromagnetic theory of gratings*. Springer, Berlin.

Reinhart, F. K., Hayashi, L. and Panish, M. B. (1971). Mode reflectivity and waveguide properties of double-heterostructure injection lasers. *Journal of Applied Physics* **42**, 4466–79.

Ryshik, I. M. and Gradstein, I. S. (1957). *Tafeln-Tables*. Veb Deutsher Verlag der Wissenshaften, Berlin.

Sameth, S. and Yariv, A. (1972). Phase matching for periodic modulation of the non-linear optical properties. *Optics Communications* **6**, 301–4.

Sarid, P. and Stegeman, G. I. (1981). Optimization of the effects of power dependent refractive index in optical waveguides. *Journal of Applied Physics* **52**, 5439–41.

Shvartsburg, A. B. (1982). Non-linear dynamics and geometric properties of localized wave structures. *Physics Reports* **83**, 109–49.

Smith, P. W., Turner, E. H., and Maloney, P. J. (1978). Electro-optic non-linear. Fabry–Perot devices. *IEEE Journal of Quantum Electronics* **QE-14**, 207–10.

Snyder, A. W. and Love, J. D. (1983). *Optical waveguide theory*. Chapman and Hall, London.

Spielmann, Ch., Krausz, F., Brabec, T., Wintner, E., and Schmidt A. J. (1991). Femtosecond passive mode locking of a solid-state laser by a dispersively balanced non-linear interferometer. *Applied Physics Letters* **58**, 2470–2.

Stegeman, G. I., Wright, E. M., Finlayson, N., Zanoni, R., and Seaton, C. T. (1988). Third-order non-linear integrated optics. *Journal of Lightwave Technology* **6**, 953–70.

Stolen, R. H. and Tom, H. W. (1987). Self-organized phase-matched harmonic generation in optical fibers. *Optics Letters* **12**, 585–7.

Tolminson, W. J. (1980). Surface waves at a non-linear interface. *Optics Letters* **5**, 323–5.

Usunov, I. M. (1990). Stationary periodical waves in optical fibers. *Optics Communications* **79**, 23–5.

Van der Ziel, I. P., Ilegems, M., Foy, P. W., and Mikulyak, R. M. (1976). Phase-matched second harmonic in a periodic GaAs waveguide. *Applied Physics Letters* **29**, 775–7.

Wabnitz, S., Wright, E. M., Seaton, C. T., and Stegeman, G. I. (1986). Instabilities and all-optical phase-controlled switching in a non-linear dielectric coherent coupler. *Applied Physics Letters* **49**, 838–40.

Webjorn, I., Laurell, F., and Arvidsson, G. (1989). Blue light generated by frequency doubling of laser diode light in a lithium niobate channel waveguide. *IEEE Photonics Technology Letters* **1**, 316–8.

Winful, H. G. (1985). Self-induced polarization changes in birefringent optical fibers. *Applied Physics Letters* **47**, 213–5.

Zakharov, V. E., Manakov, S. V., Novikov, S. P., and Pitaevsky, L. P. (1980). *The theory of solitons* [in Russian]. Nauka, Moscow.

Zeeger, K. (1973) *Semiconductor physics*. Springer-Verlag, Wien, New York.

Index